BG Chemie

Springer

Berlin
Heidelberg
New York
Barcelona
Budapest
Hong Kong
London
Milan
Paris
Tokyo

Toxicological Evaluations

9 Potential Health Hazards of Existing Chemicals

Springer

Berufsgenossenschaft
der chemischen Industrie
Kurfürsten Anlage 62

D-69115 Heidelberg

ISBN-13:978-3-642-85204-6   e-ISBN-13:978-3-642-85202-2
DOI: 10.1007/978-3-642-85202-2

Library of Congress Cataloging-in-Publication Data (Revised for vol. 9)
Toxicological evaluations.
Includes bibliographical references and indexes.
1. Toxicity testing. I. Berufsgenossenschaft der Chemischen Industrie. [DNLM: 1. Hazardous
Substances Toxicity. WA 455 T7549
RA1199.T678; 1990; 615.9'07; 90-10020
ISBN-13:978-3-642-85204-6 (Berlin)

Production: PRODUserv Springer Produktions-Gesellschaft, Berlin
Typesetting: Dataconversion by Lewis & Leins, Berlin

SPIN 10476342   51/3020 - 5 4 3 2 1 0 - Printed on acid-free paper

Preface

As part of its "Programme for the prevention of health hazards caused by industrial substances", the Berufsgenossenschaft der chemischen Industrie (BG Chemie, Employment Accident Insurance Fund of the Chemical Industry) began in 1977 to investigate the toxicity of those chemicals which are widely used, have many different applications and are suspected of being dangerous to health, in particular of having long-term effects. The investigations consist of a literature search and – depending on the results – commissions of experimental studies. It is hoped by means of this testing to close gaps in our knowledge and to increase the scientific validity of the required risk assessments. The results of the toxicological investigations carried out by BG Chemie, and the resulting substance assessments have been published in German since 1987 in the form of 169 "Toxikologische Bewertungen" ("Toxicological Evaluations") up to now.

In order to make this useful information internationally available, BG Chemie began in October 1990 to publish them as a book series in English, of which the ninth volume (containing 14 individual evaluations) is presented here. Therefore for 124 existing chemicals, "Toxicological Evaluations" are available in English at the moment, a further 41 are in preparation and will be published soon.

Because of the short time between publishing volume 1-9, printing of an "Introduction" (consisting of a general overview of the programme, lists with names of people involved as well as substances under investigation) has been abandoned in this volume and will be included in volume 10 again. If more detailed information is required about the ongoing work, see volume 1 or 4 or contact BG Chemie at first hand.

BG Chemie hopes that, for many people working in the chemical industry, this information will be of practical help in assessing hazards to health at the workplace.

Contents

Dichlorodimethylsilane

Toxicological Evaluations are also available for the structurally related compounds trichlorophenylsilane and trichloromethylsilane.

1.   Summary and assessment

Dichlorodimethylsilane is of moderate acute toxicity to rats when administered by stomach tube (oral LD_{50} 595 mg/kg body weight) and on inhalation exposure (4-hour LC_{50} 4.91 mg/l, 1-hour LC_{50} 12.5 mg/l or in the range of 16 to 25 mg/l, respectively). Signs of toxicity include ruffled fur, increased salivation, general malaise, breathing difficulties and sedation. The rats that died showed effects on the gastrointestinal tract, pancreas, liver and kidneys, and after inhalation exposure lung damage. The vapour is of high acute toxicity to mice (2-hour LC_{50} 0.3 mg/l), these much lower concentrations resulting in mucosal irritation, breathlessness, agitated motor activity, spreading of hindlimbs and laying on the stomach; the mice that died showed various changes in the lungs, liver, kidneys, spleen, heart and brain.

Application of dichlorodimethylsilane to the skin or eyes of rabbits produces severe irritation or corrosion, and the vapour is also irritating to the eyes of mice, rats and rabbits.

Both with and without metabolic activation dichlorodimethylsilane showed no mutagenic activity in the Salmonella/microsome assay and no evidence of DNA-damage in *Escherichia coli*. The substance did not induce mitotic gene conversion in *Saccharomyces cerevisiae*; in mouse lymphoma cells dichlorodimethylsilane did not induce gene mutations, sister chromatid exchange or DNA single strand breaks. In an in vitro chromosomal aberration assay the substance was clastogenic both in the presence and absence of a metabolic activation system. In an in vivo chromosomal aberration assay, however, dichlorodimethylsilane did not induce chromosomal aberrations in the bone marrow cells of rats. Considering the negative results in all relevant studies the substance is regarded to have no genotoxic potential.

2.   Name of substance

2.1	Usual name	Dichlorodimethylsilane
2.2	IUPAC-name	Dichlorodimethylsilane
2.3	CAS-No.	75–78–5
2.4	EINECS-No.	200–901–0

3.   Synonyms, common and trade names

Dichlorodimethylsilicon
Dimethyldichlorosilane
Dimethylsilane dichloride
Dimethylsilicon dichloride
Inerton AW-DMCS
Inerton DW-DMC
Repel-Silane

4.   Structural and molecular formulae

4.1   Structural formula

$$H_3C - \overset{\displaystyle Cl}{\underset{\displaystyle CH_3}{Si}} - Cl$$

4.2   Molecular formula $C_2H_6Cl_2Si$

5.   Physical and chemical properties

5.1   Molecular mass, g/mol   129.06

5.2   Melting point, °C   -60   (Bayer, 1991)

5.3   Boiling point, °C   70 (at 1013 hPa)   (Bayer, 1991)

5.4   Vapour pressure, hPa   147 (at 20 °C)
                                     500 (at 50 °C)   (Bayer, 1991)

5.5   Density, g/cm^3 1.07 (at 20 °C)   (Bayer, 1991)

5.6   Solubility in water   Reacts with water (Bayer, 1991)

5.7   Solubility in organic solvents Soluble in benzene and ether
                                                         (HSDB, 1992)

5.8   Solubility in fat   No information available

5.9   pH-value   No information available

5.10   Conversion factor   1 ml/m^3 (ppm) $\triangleq$ 5.27 mg/m^3
                                     1 mg/m^3 $\triangleq$ 0.19 ml/m^3 (ppm)
                                     (at 1013 hPa and 25 °C)

6.   Uses

Intermediate product in chemical syntheses (Bayer, 1991).

7.   Experimental results

7.1   Toxicokinetics and metabolism

No information available.

7.2   Acute and subacute toxicity

The acute oral LD_{50} of dichlorodimethylsilane was calculated to be
595 mg/kg body weight in rats (strain Bor: WISW (SPF Cpb), body

weight 174 g (males) and 167 g (females), 5 of each sex per group). The chemical was dissolved in paraffin oil and administered by stomach tube (10 ml/kg body weight) at doses of 125, 500, 707 and 1000 mg/kg body weight, and the animals were observed for 14 days. At the lowest dose the rats showed ruffled fur on the day of treatment but were otherwise unaffected. The rats given 500 and 707 mg/kg body weight developed ruffled fur, increased salivation and general malaise within 30 minutes of dosing, these effects lasting for up to 3 days (at 500 mg/kg) or until death (at 707 mg/kg); no deaths occurred at 500 mg/kg body weight, but all 10 rats given 707 mg/kg body weight died between 3 hours and 3 days after dosing. The top dose (1000 mg/kg body weight) resulted in a ruffled appearance, breathing difficulties and sedation, and the animals all died within an hour of treatment. On subsequent examination of the dead rats in the two higher-dose groups, the mucous membrane of the gastrointestinal tract was dark red to black, the pancreas was red, the liver showed a mottled brown colour, and white patches were seen in the left kidney. Several animals were cyanotic. The rats in the two lower-dose groups were killed at the end of the 14-day observation period; there were no remarkable macroscopic findings (Bayer, 1990).

In an early study, dichlorodimethylsilane was reported to be of moderate acute oral toxicity when administered by stomach tube (10 % in light mineral oil) to groups of 2 or 3 Wistar rats. No deaths occurred at 100 or 300 mg/kg body weight, but a dose of 1000 mg/kg body weight was lethal to both treated rats (no further details provided; Rowe et al., 1948).

When pairs of Wistar rats were given a single intraperitoneal injection of dichlorodimethylsilane at doses of 10, 30 or 100 mg/kg body weight (10 % in light mineral oil for the two lower doses, undiluted for the higher dose), marked tissue destruction was evident at the site of injection. One rat died at each of the lower two doses, and both died at the top dose. The surviving rats were killed 90 days after the injection; the lesions had healed with no cell proliferation (Rowe et al., 1948).

Dichlorodimethylsilane vapour was of moderate acute toxicity to rats, with LC_{50} values of 4.91 mg/l after a 4-hour exposure period (Marhold, 1986) and of 12.5 mg/l (Dow Corning, 1987) or 16–25 mg/l (Kolesar et al., 1987) after a 1-hour exposure period. High acute toxicity was reported in mice exposed for 2 hours, with an LC_{50} value of 0.3 mg/l (Korlyakova, 1961).

Respiratory difficulties were the main clinical signs observed in Sprague-Dawley rats (5 of each sex) exposed to high vapour concentrations (up to 25 mg/l) for 1 hour; the animals that died showed pulmonary oedema and focal haemorrhaging in the lungs (Kolesar et al., 1987).

Much lower concentrations (0.12–0.3 mg/l) were reported to produce breathlessness, mucosal irritation, agitated motor activity, spreading of hindlimbs and laying on the stomach in mice (strain and sex unspecified, 10 per group, body weight 20–23 g) exposed in a sealed chamber for 2 hours. At higher concentrations of 0.4–0.44 mg/l the mice displayed greater agitation, convulsions, asphyxia and cyanosis. The mice that died during the exposure period or in the subsequent 14-day observation period showed an excess of blood or fluid in all of the organs that were examined, i.e. the lungs, liver, kidneys, spleen, heart and brain. This was attributed to increased permeability of the blood vessel walls. Other effects included inflammation of the lungs and slight proliferative changes in the kidneys, spleen and heart (seen only in those mice that died in the first 24 hours), swelling and proliferation of the reticulo-endothelial elements in the liver and slight polymorphism of the liver cell nuclei. Marked degenerative changes in the brain were also observed (Korlyakova, 1961).

7.3   Skin and mucous membrane effects

Dichlorodimethylsilane was highly corrosive when applied to the shaven abdominal skin of rabbits (number, strain and sex not specified); application of an unspecified amount (presumably neat) for just 1 minute (after which time the site was rinsed) produced severe denaturation, with blistering and sloughing of the skin (Rowe et al., 1948).

Skin necrosis was also reported (no experimental details known) by Union Carbide (1986).

A further study was unable to determine the primary irritancy value for dichlorodimethylsilane on rabbit skin because of its corrosive action: 0.5 ml of the liquid was applied to the shaven flank of 1 male albino rabbit (strain HC:NZW, body weight 3.7 kg) for 4 hours under a semi-occluded patch; 1 hour after patch-removal, grey discoloration and induration of the application site were evident, and after 7 days the area was bloody and encrusted, with scabs still evident at the end of the 14-day observation period (Bayer, 1984 a).

3 drops of the chemical applied to the shaven skin of rabbits (number, strain and sex not specified) for 3 minutes produced hyperaemia on the 3rd day after application, followed by the development of a sore which quickly healed; 2 drops applied to the interior surface of the ear for 5 minutes produced mild hyperaemia lasting 1–2 days, followed by dryness of the skin, with full recovery by the 7th day (Korlyakova, 1961).

Moderate skin irritation has been reported in rabbits given a 24-hour application of 20 mg (no further details known; Marhold, 1986).

The instillation of a small drop (~ 0.05 ml) of dichlorodimethylsilane into the eyes of rabbits (number, strain and sex not specified) caused severe burns to the cornea and the eyelids (no further details; Rowe et al., 1948).

An eye injury grading of > 5 (presumably on a 10-point scale of increasing irritation) was reported (no experimental details known) by Union Carbide (1986).

A single drop of the chemical placed in both eyes of rabbits (number, strain and sex not specified) resulted in mucosal hyperaemia after 5 minutes and partial clouding of the cornea and photophobia after 13 minutes. No changes were seen in other ocular tissues, and corneal transparency was restored between days 3 and 15. When the experiment was repeated but with immediate rinsing of the eye after treatment, slight transient clouding of the cornea was observed (Korlyakova, 1961).

Severe eye irritation was also reported in rabbits after the ocular application of 5 mg dichlorodimethylsilane for 24 hours (no further details known; Marhold, 1986).

In studies involving exposure to dichlorodimethylsilane vapour, there were signs of mild eye irritation in mice exposed for 2 hours at vapour concentrations below 0.05 mg/l, and clear irritant effects were evident at higher concentrations (0.12–0.44 mg/l with inflammation of the mucous membranes of the eyes on the third or fourth day after exposure (Korlyakova, 1961).

Severe eye irritation was reported in rabbits exposed for 3 minutes at a vapour concentration of 5 mg/l (no further details known; Union Carbide, 1986).

Sprague-Dawley rats exposed to high concentrations (up to 25 mg/l) for 1 hour developed corneal opacity (Kolesar et al., 1987).

7.4   Sensitization

No information available.

7.5   Subchronic and chronic toxicity

No information available.

7.6   Genotoxicity

7.6.1 In vitro

Dichlorodimethylsilane was tested for mutagenic activity in the Salmonella/microsome assay with *Salmonella typhimurium* strains TA 98, TA 100, TA 1535 and TA 1537, both with and without metabolic activation (S9-mix from Aroclor 1254-induced rat liver). Dimethylformamide was used as a solvent. The chemical was tested at 20, 100, 500, 2500 and 12500 µg/plate and subsequently at 8, 40, 200, 1000 and 5000 µg/plate. Concentrations of 100 µg/plate and above were cytotoxic. At 5000 µg/plate a doubling in the number of revertants was observed without metabolic activation, which was not reproducible at concentrations of 3000, 4000, 5000, 6000 and 7000 µg/plate (Bayer, 1984 b).

Another study on *Salmonella typhimurium* strains TA 98, TA 100, TA 1535, TA 1537 and TA 1538 similarly found no indication of mutagenic properties at test concentrations of 0.001 – 5 µg/plate (in ethanol), with or without S9-mix (from Aroclor 1254-induced rat livers). The highest test concentration was cytotoxic (Isquith et al., 1988 a).

Dichlorodimethylsilane (> 99.8% pure; dissolved in acetone) was further tested on *Salmonella typhimurium* strains TA 98, TA 100, TA 1535, TA 1537 and TA 1538 and on *Escherichia coli* WP2uvrA with and without metabolic activation (S9-mix from Aroclor 1254-induced livers of male Sprague-Dawley rats). The concentrations used were 4–10000 µg/plate in a first experiment and 4–5000 µg/plate in a second experiment. Cytotoxicity occurred from 2500 µg/plate. Neither with nor without metabolic activation the substance showed mutagenic activity (Hoechst, 1987).

The substance, dissolved in acetone, was also negative in a Salmonella/microsome assay with preincubation either with or without metabolic activation (S9-mix from Aroclor 1254-induced rat or hamster liver) in strains TA 98, TA 100, TA 1535, TA 1537 and TA 1538. The concentrations used were 33 to 10000 µg/plate (NCI, 1988).

In a DNA-repair test with *Escherichia coli* W3110 (pol A$^+$) and P3078 (pol A$^-$), dichlorodimethylsilane showed no evidence of DNA damage at test concentrations of 0.001–5 µl/plate (in ethanol), either with or without S9-mix (from Aroclor 1254-induced rat livers). The highest test concentration was cytotoxic (Isquith et al., 1988 a).

Dichlorodimethylsilane did not induce mitotic gene conversion at the adenine-2 and tryptophan-5 loci in Saccharomyces cerevisiae D4, when tested with or without metabolic activation (S9-mix from Aroclor 1254-induced rat livers) at concentrations of 0.001–5 µl/plate (in ethanol). Cytotoxicity was evident at the highest test concentration (Isquith et al., 1988 a).

In mouse lymphoma (L5178Y) cells, dichlorodimethylsilane did not induce gene mutations at the thymidine kinase locus when tested up to a toxic concentration with or without S9-mix (from uninduced mouse livers); the chemical (in ethanol) was tested at five concentrations up to 0.32 µl/ml without S9-mix and up to 0.64 µl/ml with S9-mix. In the same test system, at concentrations of up to 0.64 µg/ml with or without S9-mix, the chemical did not induce sister chromatid exchange (SCE) but did induce chromosomal aberrations (a dose-related increase over at least two consecutive concentrations) both in the presence and absence of S9-mix. There was no sign of DNA damage in treated cells, as measured by alkaline elution (Isquith et.al., 1988 a).

Dichlorodimethylsilane (dissolved in DMSO) was also negative in a further test in mouse lymphoma L5178Y (TK$^+$/TK$^-$) cells. The concentrations used were between 0.72 and 1.74 µl/ml with metabolic activation (S9-mix from Aroclor 1254-induced rat livers) and between 0.14 and 0.35 µl/ml without metabolic activation (NCI, 1988).

7.6.2 In vivo

Dichlorodimethylsilane did not induce chromosomal aberrations in the bone marrow cells of male Sprague-Dawley rats (6 per group) given a single intraperitoneal injection of 0, 11, 22 or 32 mg/kg body weight. The test doses were selected on the basis of the maximum tolerated dose (not specified, but determined in a preliminary experiment). The bone marrow was sampled 6, 24 and 48 hours after the injection (Isquith et al., 1988 b).

7.7   Carcinogenicity

No information available.

7.8   Reproductive toxicity

No information available.

7.9   Effects on the immune system

No information available.

7.10 Neurotoxicity

No information available.

8.   Experience in humans

No specific health effects have been observed in ca. 150 workers of a organochlorosilane- and organosilane production plant in the last 20 to 25 years by the occupational medical department of the firm (Hüls, 1993).

No cases of accidental exposure to dichlorodimethylsilane have been identified. However, the chlorosilanes as a group are considered to be hazardous on acute exposure, as hydrochloric acid is formed on their hydrolysis (Sax, 1975).

Various databanks also report that dichlorodimethylsilane is hazardous on acute exposure. They state that the vapour cannot be tolerated even at low concentrations, that it is severely irritating to the eyes, nose and throat, and that if inhaled it will cause breathing difficulties. Contact with the liquid will result in severe burns of the eyes and skin, and ingestion causes severe burns of the mouth and stomach (CHRIS, 1991; HSDB, 1992). The basis for these statements is not known; they may simply be derived from the animal data.

9.   Threshold limit values

No information available.

References

Bayer AG, Institut für Toxikologie
Dimethyldichlorsilan - Prüfung auf primär reizende/ätzende Wirkung an der Kaninchenhaut
Unpublished report, study no. T 1019335 (1984 a)

Bayer AG, Institut für Toxikologie
Dimethyldichlorsilan - Salmonella/Mikrosomen-Test zur Untersuchung auf punktmutagene Wirkung
Unpublished report no. 12768 (1984 b)

Bayer AG, Institut für Toxikologie
Dimethyldichlorsilan - Untersuchungen zur akuten oralen Toxizität an männlichen und weiblichen Wistar-Ratten
Unpublished report no. 19021 (1990)

Bayer AG
AIDA-Grunddatensatz Silane, dichloromethyl- (1991)

CHRIS (Chemical Hazard Response Information System)
The United States Coast Guard, US Department of Transportation
SilverPlatter, Chem-Bank, March 1991

Dow Corning
Unpublished data (1987)
Cited in: Bayer (1991)

Hoechst AG, Pharma Research Toxicology and Pathology
Silan M2 - Study of the mutagenic potential in strains of *Salmonella typhimurium* (Ames test) and *Escherichia coli*
Unpublished report no. 87.0983 (1987)
on behalf of Wacker Chemie GmbH

HSDB (Hazardous Substances Data Bank)
US National Library of Medicine
SilverPlatter, Chem-Bank, Februar 1992

Hüls AG, Werksärztlicher Dienst, Werk Rheinfelden
Written communication to BG Chemie, 10.03.1993

Isquith, A., Matheson, D., Slesinski, R.
Genotoxicity studies on selected organosilicon compounds: in vitro assays
Food Chem. Toxicol., 26, 255 - 261 (1988 a)

Isquith, A., Slesinski, R., Matheson, D.
Genotoxicity studies on selected organosilicon compounds: in vivo assays
Food Chem. Toxicol., 26, 263–266 (1988 b)

Kolesar, G.B., Siddiqui, W.H., Hobbs, E.J.
A comparison of acute inhalation toxicity of a series of chlorosilanes with hydrogen chloride
Toxicologist, 7, 192, Abstract No. 768 (1987)

Korlyakova, E.A.
Toxicology of certain monomers of organosilicon compounds (trichloromethylsilane, dichlorodimethylsilane, trichloroethylsilane, trichlorophenylsilane)
Toksikol. Nov. Prom. Khim. Veshchestv., 3, 23–33 (1961)

Marhold, J.V.
Dichlor-dimethylsilan
Prehled Prumyslove Toxikol. Org. Latky, p. 1219 (1986)

NCI (National Cancer Institute)
Short-term test program sponsored by the Division of Cancer Etiology (1988)

Rowe, V.K., Spencer, H.C., Bass, S.L.
Toxicological studies on certain commercial silicones and hydrolyzable silane intermediates
J. Ind. Hyg. Toxicol., 30, 332–352 (1948)

Sax, N.I.
Dangerous properties of industrial materials
4th ed., p. 554
Van Nostrand Reinhold Company, New York (1975)

Union Carbide Corporation
Rabbit eye and skin injury testing on seven silica compounds
NTIS/OTS 0000469–0 (1986)

Trichloromethylsilane

Toxicological Evaluations are also available for the structurally related compounds trichlorophenylsilane and dichlorodimethylsilane.

1.   Summary and assessment

Trichloromethylsilane is of moderate toxicity to rats when administered orally (by stomach tube; 300 mg/kg body weight are not lethal) or on inhalation exposure (4-hour LC_{50} 2.74 mg/l, 1-hour LC_{50} 9.6 ml/l or in the range of 20 to 30 mg/l, respectively). Signs of toxicity included breathing difficulties after inhalation; macroscopically lung damage was seen after inhalation of lethal concentrations. The vapour is of high acute toxicity to mice (2-hour LC_{50} 0.18 mg/l), these much lower test concentrations resulting in mucosal irritation, breathlessness agitated motor activity, spreading of hindlimbs and laying on the stomach; the mice that died showed various changes in the lungs, liver, kidneys, spleen, heart and brain.

Trichloromethylsilane is corrosive when applied to the skin or eyes of rabbits, and the vapour is irritating to the eyes of mice and rats.

Both with and without metabolic activation trichloromethylsilane showed no mutagenic activity in the Salmonella/microsome assay and no evidence of DNA-damage in *Escherichia coli*. The substance did not induce mitotic gene conversion in *Saccharomyces cerevisiae* or gene mutations and chromosomal aberrations in mouse lymphoma cells. An increased frequency of sister chromatid exchanges was observed in mouse lymphoma cells, but no induction of DNA single strand breaks. In vivo trichloromethylsilane did not induce chromosomal aberrations in the bone marrow cells of rats given a single intraperitoneal injection. Considering the negative results in all relevant studies the substance is regarded to have no genotoxic potential.

2.   Name of substance

2.1   Usual name   Trichloromethylsilane

2.2   IUPAC-name   Methyltrichlorosilane

2.3   CAS-No.   75–79–6

2.4   EINECS-No.   202–640–8

3.   Synonyms, common and trade names

Methyl silico chloroform
Methylsilicon trichloride
Methylsilyl trichloride
Methyltrichlorsilan
Trichloromethylsilicon
Trichlormethylsilan

4.   Structural and molecular formulae

4.1   Structural formula

$$H_3C - \underset{\underset{Cl}{|}}{\overset{\overset{Cl}{|}}{Si}} - Cl$$

4.2   Molecular formula CH_3Cl_3Si

5.   Physical and chemical properties

5.1   Molecular mass, g/mol   149.48

5.2   Melting point, °C   - 90   (Bayer, 1991)

5.3   Boiling point, °C   66 (at 1013 hPa)   (Bayer, 1991)

5.4	Vapour pressure, hPa	189 (at 20 °C)	
		593 (at 50 °C)	(Bayer, 1991)
5.5	Density, g/cm^3	1.27 (at 20 °C)	(Bayer, 1991)
5.6	Solubility in water	Reacts violently with water	
			(Bayer, 1991)
5.7	Solubility in organic solvents	No information available	
5.8	Solubility in fat	No information available	
5.9	pH-value	No information available	
5.10	Conversion factor	1 ml/m^3 (ppm) $\triangleq$ 6.10 mg/m^3	
		1 mg/m^3 $\triangleq$ 0.16 ml/m^3 (ppm)	
		(at 1013 hPa and 25 °C)	

6.   Uses

Starting product for chemical syntheses; used in the manufacture of silicon resins and pyrogenic silicic acid (Bayer, 1991).

7.   Experimental results

7.1   Toxicokinetics and metabolism

No information available.

7.2   Acute and subacute toxicity

Trichloromethylsilane was reported to be of moderate acute oral toxicity when administered by stomach tube (10 % in light mineral oil) to groups of 2 or 3 Wistar rats. No deaths occurred at 100 or 300 mg/kg body weight, but a dose of 1000 mg/kg body weight was lethal to both treated rats (no further details provided; Rowe et al., 1948).

When pairs of Wistar rats were given a single intraperitoneal injection of trichloromethylsilane at doses of 10, 30, 100 or 300 mg/kg body weight (10% in light mineral oil for the two lower doses, undiluted for the two higher doses), marked tissue destruction was evident at the site of injection. Both rats survived the lowest dose, one died at each of the middle two doses, and both died at the top dose. The surviving rats were killed 90 days after the injection; the lesions had healed with no cell proliferation (Rowe et al., 1948).

Trichloromethylsilane vapour was of moderate acute toxicity to rats, with LC_{50} values of 2.74 mg/l after a 4-hour exposure period (Marhold, 1986) and of 9.6 mg/l (Dow Corning, 1987) or 20–30 mg/l (Kolesar et al., 1987) after a 1-hour exposure period. High acute toxicity was reported in mice exposed for 2 hours, with an LC_{50} value of 0.18 mg/l (Korlyakova, 1961).

Respiratory difficulties were the main clinical signs observed in Sprague-Dawley rats (5 of each sex) exposed to high vapour concentrations (up to 30 mg/l) for 1 hour in a 3-litre nose-only chamber; the animals that died showed pulmonary oedema and focal haemorrhaging in the lungs (Kolesar et al., 1987).

Much lower concentrations (0.12–0.44 mg/l) were reported to produce breathlessness in mice (strain and sex unspecified, 10 per group, body weight 20–23 g) exposed in a sealed chamber; mucosal irritation, agitated motor activity, spreading of hindlimbs, laying on the stomach, convulsions, asphyxia and cyanosis were also reported. The mice that died during the exposure period or in the subsequent 14-day observation period showed an excess of blood or fluid in all of the organs that were examined, i.e. the lungs, liver, kidneys, spleen, heart and brain. This was attributed to increased permeability of the blood vessel walls. Other effects included inflammation of the lungs and slight proliferative changes in the kidneys, spleen and heart (seen only in those mice that died in the first 24 hours), swelling and proliferation of the reticulo-endothelial elements in the liver and slight polymorphism of the liver cell nuclei. Degenerative changes in the brain were also observed (Korlyakova, 1961).

7.3   Skin and mucous membrane effects

Trichloromethylsilane was highly corrosive when applied to the shaven abdominal skin of rabbits (number, strain and sex not specified); application of an unspecified amount (presumably neat) for just 2–3 minutes (after which time the site was rinsed) produced severe denaturation, with blistering and sloughing of the skin (Rowe et al., 1948).

3 drops of the chemical applied to the shaven skin of rabbits (number, strain and sex not specified) for 3 minutes produced hyperaemia within 1 or 2 days, followed by the development of a sore which healed fairly quickly; 2 drops applied to the interior surface of the ear for 5 minutes produced mild hyperaemia lasting 1–2 days, followed by dryness of the skin, with full recovery by the 7th day (Korlyakova, 1961).

Mild skin irritation has been reported in rabbits given a 24-hour application of 20 mg (no further details known; Marhold, 1986).

The instillation of a small drop (~ 0.05 ml) of trichloromethylsilane into the eyes of rabbits (number, strain and sex not specified) caused severe burns to the cornea and the eyelids (no further details; Rowe et al., 1948).

Similarly, a single drop of the chemical placed on the cornea of rabbits (number, strain and sex not specified) provoked clear conjunctival irritation, with marked clouding of the cornea, epithelial damage, mucosal discharge and necrosis of the mucous membrane of the lower eyelid; ulceration of the eyelids was evident on the fifth day after application (Korlyakova, 1961).

Severe eye irritation was also reported in rabbits after the ocular application of 5 mg trichloromethylsilane for 24 hours (no further details known; Marhold, 1986).

In studies involving exposure to trichloromethylsilane vapour, signs of mild eye irritation were reported in mice exposed for 2 hours at vapour concentrations below 0.05 mg/l, and clear irritant effects were evident at higher concentrations (0.12–0.44 mg/l) with inflammation of the mucous membranes of the eyes on the third or fourth day after exposure (Korlyakova, 1961).

Corneal opacity was reported in Sprague-Dawley rats exposed to high concentrations (up to 30 mg/l) for 1 hour (Kolesar et al., 1987).

7.4   Sensitization

No information available.

7.5   Subchronic and chronic toxicity

No information available.

7.6   Genotoxicity

7.6.1 In vitro

Trichloromethylsilane was tested for mutagenic activity in the Salmonella/microsome assay with *Salmonella typhimurium* strains TA 98, TA 100, TA 1535 and TA 1537, both with and without metabolic activation (S9-mix from Aroclor 1254-induced rat liver). Ethylene glycol dimethyl ether was used as a solvent. In the first experiment, all test concentrations (20, 100, 500, 2500 and 12 500 µg/plate) had a cytotoxic effect, so the test was repeated at 75, 150, 300, 600, 1200 and 2400 µg/plate. There was no evidence of mutagenic activity at any concentration, either with or without S9-mix (Bayer, 1988).

Another study on *Salmonella typhimurium* strains TA 98, TA 100, TA 1535, TA 1537 and TA 1538 similarly found no indication of mutagenic properties at test concentrations of 0.001 – 5 µl/plate (in ethanol), with or without S9-mix (from Aroclor 1254-induced rat livers). The highest test concentration was cytotoxic (Isquith et al., 1988 a).

In a DNA-repair test with *Escherichia coli* W3110 (pol A$^+$) and P3078 (pol A$^-$), trichloromethylsilane showed no (indirect) evidence of DNA damage at test concentrations of 0.001 – 5 µl/plate (in ethanol), either with or without S9-mix (from Aroclor 1254-induced rat livers). The highest test concentration was cytotoxic (Isquith et al., 1988 a).

Trichloromethylsilane did not induce mitotic gene conversion at the adenine-2 and tryptophan-5 loci in *Saccharomyces cerevisiae* D4, when tested with or without metabolic activation (S9-mix from Aroclor 1254-induced rat livers) at concentrations of 0.001 – 5 µg/-plate (in ethanol). Cytotoxicity was evident at the highest test concentration (Isquith et al., 1988 a).

In mouse lymphoma (L5178Y) cells, trichloromethylsilane did not induce gene mutations at the thymidine kinase locus when tested up to a toxic concentration with or without S9-mix (from uninduced mouse livers); the chemical (in ethanol) was tested at five concentrations up to 0.32 µl/ml. In the same test system the chemical did induce sister chromatid exchange (SCE) when tested at concentrations of up to 0.16 µg/ml without S9-mix and up to 0.32 µg/ml with S9-mix. The increased frequency of SCEs was dose-related over at least two consecutive doses. The same test concentrations did not produce any evidence of chromosomal aberrations. Similarly, there was no sign of DNA damage in treated cells, as measured by alkaline elution (Isquith et al., 1988 a).

7.6.2 In vivo

Trichloromethylsilane did not induce chromosomal aberrations in the bone marrow cells of male Sprague-Dawley rats (6 per group) given a single intraperitoneal injection of 0, 12, 16 or 23 mg/kg body weight. The test doses were selected on the basis of the maximum tolerated dose (not specified, but determined in a preliminary experiment). The bone marrow was sampled 6, 24 and 48 hours after the injection (Isquith et al., 1988 b).

7.7   Carcinogenicity

No information available.

7.8   Reproductive toxicity

No information available.

7.9   Effects on the immune system

No information available.

7.10 Neurotoxicity

No information available.

8.   Experience in humans

No specific health effects have been observed in ca. 150 workers of a organochlorosilane- and organosilane production plant in the last

20 to 25 years by the occupational medical department of the firm (Hüls, 1993).

No cases of accidental exposure to trichloromethylsilane have been identified. However, the chlorosilanes as a group are considered to be hazardous on acute exposure, as hydrochloric acid is formed on their hydrolysis (Sax, 1975).

Various databanks also report that trichloromethylsilane is hazardous on acute exposure. They state that the vapour cannot be tolerated even at low concentrations, that it is severely irritating to the eyes, nose and throat, and that if inhaled it will cause breathing difficulties. Contact with the liquid will results in severe burns of the eyes and skin, and ingestion causes severe burns of the mouth and stomach (CHRIS, 1991; HSDB, 1992). The basis for these statements is not known; they may simply be derived from the animal data.

9.   Threshold limit values

No information available.

References

Bayer AG, Fachbereich Toxicology
Trichloromethylsilane - Salmonella/microsome test to evaluate for point mutagenic effects
Unpublished report no. 17233 (1988)

Bayer AG
AIDA-Grunddatensatz Silane, trichloromethyl- (1991)

CHRIS (Chemical Hazard Response Information System)
The United States Coast Guard, US Department of Transportation
SilverPlatter, Chem-Bank (1991)

Dow Corning
Unpublished data (1987)
Cited in: Bayer (1991)

HSDB (Hazardous Substances Data Bank)
US National Library of Medicine
SilverPlatter, Chem-Bank (1992)

Hüls AG, Werksärztlicher Dienst, Werk Rheinfelden
Written communication to BG Chemie, 10.03.1993

Isquith, A., Matheson, D., Slesinski, R.
Genotoxicity studies on selected organosilicon compounds: in vitro assays
Food Chem. Toxicol., 26, 255–261 (1988 a)

Isquith, A., Slesinski, R., Matheson, D.
Genotoxicity studies on selected organosilicon compounds: in vivo assays
Food Chem. Toxicol., 26, 263–266 (1988 b)

Kolesar, G.B., Siddiqui, W.H., Hobbs, E.J.
A comparison of acute inhalation toxicity of a series of chlorosilanes with hydrogen chloride
Toxicologist, 7, 192, Abstract No. 768 (1987)

Korlyakova, E.A.
Toxicology of certain monomers of organosilicon compounds (trichloromethylsilane, dichlorodimethylsilane, trichloroethylsilane, trichlorophenylsilane)
Toksikol. Nov. Prom. Khim. Veshchestv., 3, 23–33 (1961)

Marhold, J.V.
Trichlor-methylsilan
Prehled Prumyslove Toxikol. Org. Latky, p. 1219 (1986)   .

Rowe, V.K., Spencer, H.C., Bass, S.L.
Toxicological studies on certain commercial silicones and hydro-

lyzable silane intermediates
J. Ind. Hyg. Toxicol., 30, 332–352 (1948)

Sax, N.I.
Dangerous properties of industrial materials
4th ed., p. 554
Van Nostrand Reinhold Company, New York (1975)

2-Nitro-1,3-dimethylbenzene

1. Summary and assessment

2-Nitro-1,3-dimethylbenzene is of very low acute toxicity (LD_{50} rat oral, males 3330 mg/kg body weight, females 2630 mg/kg body weight).

Repeated administration in the diet of 0, 200, 1000 or 5000 ppm 2-nitro-1,3-dimethylbenzene for 5 days or of 0, 100, 600 or 3000 ppm for 28 days (equivalent to 0, 20, 100 and 500 mg/kg body weight/day and 0, 10, 60 and 300 mg/kg body weight/day, respectively), led to an increase in the relative liver weight at 5000 ppm in the 5-day study. In the 28-day study, an increase in the activity of γ-glutamyl transferase and an increase in the relative liver weight were seen at 3000 ppm. These effects were reversible within 14 days. No treatment-related effects were detectable histologically in the liver, kidneys, adrenal glands, spleen or testes. The no effect level reported in this study is 600 ppm (ca. 60 mg/kg body weight/day).

2-Nitro-1,3-dimethylbenzene causes only very mild, transient erythema when applied semi-occlusively to rabbit skin for 4 hours. The substance has been evaluated as not irritating in accordance with the guidelines in 83/467/EEC. Instillation into the eye causes reddening of the conjunctiva lasting only a few hours, which clears up within 24 hours. In accordance with EC classification criteria, 2-nitro-1,3-dimethylbenzene has also been evaluated as not irritating to the eye.

2-Nitro-1,3-dimethylbenzene is not mutagenic in the Salmonella/microsome test either with or without S9-mix. Structural chromosome aberrations are induced in the presence of S9-mix in the chromosome aberration test in V79 cells in vitro.

2. Name of substance

2.1   Usual name   2-Nitro-1,3-dimethylbenzene

2.2	IUPAC-name	2-Nitro-1,3-dimethylbenzene
2.3	CAS-No.	81–20–9
2.4	EINECS-No.	201–333–6

3.   Synonyms, common and trade names

1,3-Dimethyl-2-nitrobenzene
2-Nitro-1,3-dimethylbenzol
2,6-Dimethylnitrobenzene
2-Nitro-m-xylene
Nitro-m-xylene (213)
Nitro-m-xylene, vic.

4.   Structural and molecular formulae

4.1   Structural formula

$$CH_3$$ NO_2 CH_3

4.2   Molecular formula $C_8H_9NO_2$

5.   Physical and chemical properties

5.1	Molecular mass, g/mol	151.16
5.2	Melting point, °C	15 (Thiem et al., 1979)
5.3	Boiling point, °C	222 (at 1000 hPa) (Thiem et al., 1979)
5.4	Vapour pressure, hPa	No information available
5.5	Density, g/cm³	1.112 (at 15 °C) (Thiem et al., 1979)

5.6	Solubility in water	Insoluble (Thiem et al., 1979)
5.7	Solubility in organic solvents	Dissolves readily in most organic solvents (Thiem et al., 1979)
5.8	Solubility in fat	No information available
5.9	pH-value	-
5.10	Conversion factor	$1\ ml/m^3$ (ppm) $\triangleq 6.17\ mg/m^3$ $1\ mg/m^3 \triangleq 0.16\ ml/m^3$ (ppm) (at 1013 hPa and 25 °C)

6.   Uses

In the manufacture of 2,6-xylidine, from which dyestuffs, pharmaceuticals, herbicides and fungicides are manufactured (Thiem et al., 1979).

7.   Experimental results

7.1   Toxicokinetics and metabolism

No information available.

7.2   Acute and subacute toxicity

The acute oral toxicity of 2-nitro-1,3-dimethylbenzene was studied in Wistar rats (5 males and 5 females/dose, 14-day observation period). LD_{50} values of 3330 (2920 to 3680) mg/kg body weight in males and 2630 (2310 to 2900) mg/kg body weight in females were determined. Signs of toxicity included impairment of breathing, of the locomotor system, of consciousness and of the reflexes, as well as yellow coloration of the urine. The signs of toxicity were reversible from day 5 of the observation period. In the animals that died during the study, congestion of the lungs and discoloration of the liver, adrenal glands and spleen were seen at post-mortem. No macroscopic effects were evident in the rats killed at the end of the study (Hoechst, 1986 a).

In a range-finding study for a 28-day feeding study, groups of 5 male and 5 female Wistar rats aged 3 to 4 weeks were given 0, 200, 1000 or 5000 ppm 2-nitro-1,3-dimethylbenzene in the diet for 5 days (purity 99.9 %; equivalent to 0, 20, 100 and 500 mg/kg body weight/day, respectively). The homogeneous distribution of the test substance and its stability in the feed complied with requirements. General behaviour, weight gain, feed intake and haematology (red and white blood counts) were unaffected. The relative liver weight was increased in both sexes at the top dose only. No treatment-related effects were found at autopsy (TNO, 1991).

In the subsequent 28-day feeding study, which was carried out in accordance with OECD guideline no. 407, concentrations of 0, 100, 600 and 3000 ppm 2-nitro-1,3-dimethylbenzene in the diet were chosen (purity 99.9 %; equivalent to 0, 10, 60 and 300 mg/kg body weight/day, respectively). The homogeneity and stability of the test substance in the feed complied with requirements. Groups of 5 male and 5 female rats/dose were used, with a further 5 male and 5 female rats at 0 and 3000 ppm being kept for a 14-day observation period. There were no differences in general behaviour, body weight gain, feed intake or haematology (haemoglobin, erythrocytes, leukocytes, differential blood count, prothrombin time, thrombocytes) compared with the controls. The activity of γ-glutamyl transferase was increased at the end of the treatment period, but had returned to normal by the end of the observation period. Urinalysis revealed no differences between the treated and control rats. At the end of the treatment period, the relative liver weights at the highest dose level were statistically significantly increased in the males and clearly but not statistically significantly increased in the females. This effect was no longer evident at the end of the observation period. No treatment-related effects were found on histological examination of the liver, kidneys, heart, adrenal glands, spleen or testes. The no effect level reported for this study was 600 ppm (ca. 60 mg/kg body weight/day; TNO, 1991).

7.3   Skin and mucous membrane effects

The acute skin irritancy of 2-nitro-1,3-dimethylbenzene (> 99 % pure) was studied after application of the substance to the clipped dorsal skin of New Zealand albino rabbits in accordance with

OECD guideline no. 404 (4-hour semi-occlusive exposure to the neat substance). The skin was free from signs of irritation up to 48 hours after removal of the patch, while after 72 hours, a very slight, clear, localized erythema was seen in 2 of 3 rabbits, which was still evident in one of the rabbits after 7 days. According to the classification criteria used (the guidelines in 83/467/EEC), the substance was evaluated as not irritating (Hoechst, 1986 b).

The acute eye irritancy of > 99 % pure 2-nitro-1,3-dimethylbenzene was studied in New Zealand albino rabbits in accordance with OECD guideline no. 405 (0.1 ml/eye, neat). One hour after instillation, the conjunctival blood vessels of all of the rabbits showed marked hyperaemia, and in 2 of the 3 rabbits a clear discharge was seen. None of the rabbits showed irritant effects 24 hours after instillation. The product was evaluated as not irritating (guidelines in 83/467/EEC; Hoechst, 1986 c).

7.4   Sensitization

No information available.

7.5   Subchronic and chronic toxicity

No information available.

7.6   Genotoxicity

7.6.1 In vitro

2-Nitro-1,3-dimethylbenzene (99.9 % pure) was tested in the plate incorporation test with and without S9-mix (from Aroclor 1254-induced rat liver) in the *Salmonella typhimurium* strains TA 1535, TA 1537, TA 1538, TA 98 and TA 100 and in *Escherichia coli* WP2. Concentrations of from 3.3 to 5000 µg/plate were used. The substance was not mutagenic either with or without S9-mix. Bacterial growth was reduced from 333.3 µg/plate (CCR, 1988 a).

No mutagenic activity was observed in an Ames test in strains TA 98 and TA 100 with or without S9-mix (from Kanechlor KC-400-induced rat liver; no further details; Nohmi et al., 1984).

A further publication similarly reported the absence of mutagenic activity in strains TA 100 and TA 98 (no further details; Kawai et al., 1987).

2-Nitro-1,3-dimethylbenzene (99.9 % pure) was studied in a chromosome aberration test in Chinese hamster V79 cells in vitro with and without S9-mix (from Aroclor 1254-induced rat liver). Slides were prepared 7, 18 and 28 hours after the beginning of the study. A concentration of 100 µg/ml was tested at all three time points and concentrations of 5 and 50 µg/ml were also tested at the 18-hour time point. In the presence of S9-mix, 100 µg 2-nitro-1,3-dimethylbenzene/ml induced structural chromosome aberrations. 2-Nitro-1,3-dimethylbenzene was thus mutagenic in this test system (CCR, 1988 b).

7.6.2 In vivo

No information available.

7.7   Carcinogenicity

No information available.

7.8   Reproductive toxicity

No information available.

7.9   Effects on the immune system

No information available.

7.10 Neurotoxicity

No information available.

7.11 Other effects

No information available.

8.   Experience in humans

No information available.

9.   Threshold limit values

No information available.

References

CCR (Cytotest Cell Research GmbH & Co KG)
Salmonella typhimurium and *Escherichia coli* reverse mutation assay with 2-Nitro-1,3-dimethylbenzol
Unpublished report, CCR project 116010 (1988 a)
On behalf of BG Chemie

CCR (Cytotest Cell Research GmbH & Co KG)
Chromosome aberration assay in Chinese hamster V79 cells in vitro with 2-Nitro-1,3-dimethylbenzol
Unpublished report, CCR project 116021 (1988 b)
On behalf of BG Chemie

Hoechst AG, Pharma Forschung Toxikologie
Nitro-m-xylol vic. - Prüfung der akuten oralen Toxizität an der männlichen und weiblichen Wistar-Ratte
Unpublished report no. 86.0471 (1986 a)

Hoechst AG, Pharma Forschung Toxikologie
Nitro-m-xylol vic. - Prüfung auf Hautreizung am Kaninchen
Unpublished report no. 86.0453 (1986 b)

Hoechst AG, Pharma Forschung Toxikologie
Nitro-m-xylol vic. - Prüfung auf Augenreizung am Kaninchen
Unpublished report no. 86.0397 (1986 c)

Kawai, A., Goto, S., Matsumoto, Y., Matsushita, H.
Mutagenicity of aliphatic and aromatic nitro compounds
Jpn. J. Ind. Health (Sangyo Igaku), 29, 34–54 (1987)

Nohmi, T., Yoshikawa, K., Nakadate, M., Miyata, R., Ishidate, M., jr.
Mutations in *Salmonella typhimurium* and inactivation of *Bacillus subtilis* transforming DNA induced by phenylhydroxylamine derivatives
Mutat. Res., 136, 159–168 (1984)

Thiem, K.W., Sewekow, B., Kiel, W., Handschuh, V., Freese, H., Schimpf, R., Vagt, H., Bunge, W.
Nitroverbindungen, aromatische
In: Ullmanns Encyklopädie der technischen Chemie
4th ed., vol. 17, p. 383–416
Verlag Chemie, Weinheim (1979)

TNO Nutrition and Food Research, Zeist, Netherlands
Short-term oral toxicity study with 2-Nitro-1,3-dimethylbenzol in rats: - range-finding (5-day) study, - sub-acute (28-day) study with a (14-day) recovery study
Unpublished report no. V88.463 (1991)
On behalf of BG Chemie

3-Nitro-1,2-dimethylbenzene

1.   Summary and assessment ____________________________

3-Nitro-1,2-dimethylbenzene is of low acute toxicity on oral administration (LD_{50} rat oral, male 2380 mg/kg body weight, female 2110 mg/kg body weight).

On repeated administration in the diet of 0, 200, 1000 or 5000 ppm for 5 days or of 0, 100, 500 or 2500 ppm for 28 days (equivalent to 0, 20, 100 and 500 mg/kg body weight/day and 0, 10, 50 and 250 mg/kg body weight/day, respectively), reduced weight gain and reduced feed intake were found in both sexes at 5000 ppm in the range-finding study and at 2500 ppm in the main study. After administration for 5 days, the relative liver weight was dose-dependently increased in the males in all treated groups. No effects were found at post-mortem. Functional disturbance of the liver were found in males and females after administration of 2500 ppm for 4 weeks (decrease in total protein and effects on alanine aminotransferase activity and the albumin-globulin quotient). However, no effects on liver weight or histologically detectable damage to the liver were seen. The no toxic effect level derived from the 28-day study is 500 ppm (equivalent to ca. 50 mg/kg body weight/day).

Exposure of rabbit skin to 3-nitro-1,2-dimethylbenzene leads to slight, transient inflammatory irritation. Similarly, the substance causes transient inflammatory effects on the conjunctiva and iris of the eye. The product has been evaluated as not irritating in accordance with EC classification criteria in both studies.

3-Nitro-1,2-dimethylbenzene is not mutagenic in the Salmonella/microsome test. In the chromosome aberration test in V79 cells in vitro, 3-nitro-1,2-dimethylbenzene induces structural chromosome aberrations.

2.   Name of substance

2.1	Usual name	3-Nitro-1,2-dimethylbenzene
2.2	IUPAC-name	3-Nitro-1,2-dimethylbenzene
2.3	CAS-No.	83–41–0
2.4	EINECS-No.	201–474–3

3.   Synonyms, common and trade names

1,2-Dimethyl-3-nitrobenzene
3-Nitro-1,2-dimethylbenzol
2,3-Dimethylnitrobenzene
3-Nitro-o-xylene
Nitro-o-xylene (312)
Nitro-o-xylene, vic.

4.   Structural and molecular formulae

4.1   Structural formula

4.2   Molecular formula $C_8H_9NO_2$

5.   Physical and chemical properties

5.1	Molecular mass, g/mol	151.16
5.2	Melting point, °C	15 (Thiem et al., 1979)
5.3	Boiling point, °C	246 (at 1013 hPa) 116 (at 13 hPa) (Thiem et al., 1979)
5.4	Vapour pressure, hPa	No information available

5.5	Density, g/cm^3	1.1402 (at 20 °C) (Thiem et al., 1979)
5.6	Solubility in water	Insoluble (Thiem et al., 1979)
5.7	Solubility in organic solvents	Dissolves readily in most organic solvents (Thiem et al., 1979)
5.8	Solubility in fat	No information available
5.9	pH-value	–
5.10	Conversion factor	1 ml/m^3 (ppm) $\triangleq$ 6.17 mg/m^3 1 mg/m^3 $\triangleq$ 0.16 ml/m^3 (ppm) (at 1013 hPa and 25 °C)

6.   Uses

In the manufacture of 2,3-xylidine, from which azo dyes are manufactured (Thiem et al., 1979)

7.   Experimental results

7.1   Toxicokinetics and metabolism

No information available.

7.2   Acute and subacute toxicity

The acute oral toxicity of 3-nitro-1,2-dimethylbenzene was studied in Wistar rats (5 males and 5 females/dose, 14-day observation period). The LD_{50} value in males was 2380 mg/kg body weight, while that in females was 2110 mg/kg body weight. Signs of toxicity observed included impairment of breathing, coordination and reflexes, as well as stupefaction and blood-coloured encrustation of the muzzles. In the males and females that died during the study, post-mortems revealed that the livers and spleens were pale-coloured, and the small intestine was filled with milky to orange-coloured, blood-containing (haemoccult test) fluid. No pathological effects were found on the internal organs of the rats killed at the end of the observation period (Hoechst, 1987 a).

In a 5-day range-finding study, groups of 5 male and 5 female Wistar rats, aged 3 to 4 weeks were given 0, 200, 1000 or 5000 ppm 3-nitro-1,2-dimethylbenzene in the diet (purity 99.9%; equivalent to 0, 20, 100 and 500 mg/kg body weight/day, respectively). Storage of the feed for 4 or 7 days in open containers at room temperature led to a slight, but acceptable, loss of the substance (maximum 24 % after 7 days). The distribution was homogeneous. Body weight gain and feed intake were reduced in both sexes in the 5000 ppm group. There were no effects on haematological parameters in any of the test groups. The relative liver weights were dose-dependently increased in the males in all test groups. The relative spleen weights were reduced in the females in the 5000 ppm group. No effects were found at autopsy (TNO, 1991).

Concentrations of 0, 100, 500 and 2500 ppm 3-nitro-1,2-dimethylbenzene (99.9 % pure) in the diet (equivalent to 10, 50 and 250 mg/kg body weight/day, respectively) were chosen for the subsequent 28-day feeding study which was carried out in accordance with OECD guideline no. 407. 5 male and 5 female rats/dose were used, and a further 5 male and 5 female rats were included at 0 and 2500 ppm and kept for a 14-day observation period. No changes in general behaviour or signs of toxicity were seen at any time. A reduction in body weight and feed intake in both sexes occurred at the high dose level (2500 ppm). In addition, a reduction in total protein (both sexes) and in alanine aminotransferase activity (-females) and an increase in the albumin-globulin quotient (males) and the γ-glutamyl transferase activity (females) were seen. A reduction in total protein in the females and an increase in the albumin-globulin quotient in the males in the medium-dose group was not thought to be of any toxicological significance. Compared with the controls, no changes in weight or histopathology of the liver were detectable. There were also no effects on the weights of the adrenal glands, kidneys, testes or spleen. There were no treatment-related effects on any of the organs examined histologically (adrenal glands, heart, kidneys, liver, spleen) either at the end of the study or at the end of the observation period. The no toxic effect level determined in this 28-day study was 500 ppm in the diet (equivalent to ca. 50 mg/kg body weight/day; TNO, 1991).

7.3   Skin and mucous membrane effects

The acute skin irritancy of 3-nitro-1,2-dimethylbenzene (> 99% pure) was studied after application of the substance to the clipped dorsal skin of New Zealand albino rabbits in accordance with OECD guideline no. 404 (4-hour semi-occlusive exposure to the neat substance). 1 to 24 hours after removal of the patch, barely perceptible to clear localized erythema and barely perceptible to clear localized oedema occurred. All of the irritant effects had cleared up 72 hours after removal of the patch. In accordance with the classification criteria in the guidelines contained in 83/467/EEC, the product was evaluated as not irritating (Hoechst, 1987 b).

The acute eye irritancy of > 99 % pure 3-nitro-1,2-dimethylbenzene was studied in New Zealand albino rabbits in accordance with OECD guideline no. 405 (0.1 ml/eye, neat). One hour after instillation, the conjunctiva showed diffuse crimson to severe red coloration accompanied by mild to marked conjunctival swelling. The iris was reddened. The irritant effects were reversible from 48-hours. The product was evaluated as not irritating to the eye in accordance with the classification criteria in the guidelines in 83/467/EEC (Hoechst, 1987 c).

7.4   Sensitization

No information available.

7.5   Subchronic and chronic toxicity

No information available.

7.6   Genotoxicity

7.6.1 In vitro

3-Nitro-1,2-dimethylbenzene (> 99.9 % pure) was tested in the Salmonella/microsome test in strains TA1535, TA1537, TA1538,

TA 98 and TA 100 and in *Escherichia coli* WP2. Concentrations of from 10 to 5000 µg/plate were used with and without S9-mix (from Aroclor 1254-induced rat liver). Clear indications of bacteriotoxicity were seen at 1000 and 5000 µg/plate. No increase in the revertant count was found in any of the strains, either with or without the addition of S9-mix (CCR, 1988 a).

No mutagenic activity was observed in two further studies in strains TA 98 and TA 100 with and without the addition of S9-mix (from Kanechlor KC-400- and 500-induced rat liver, respectively; no further details; Nohmi et al., 1984; Kawai et al., 1987).

3-Nitro-1,2-dimethylbenzene (99.9 % pure) was studied in a chromosome aberration test in Chinese hamster V79 cells in vitro with and without S9-mix (from Aroclor 1254-induced rat liver). Slides were prepared 7, 18 and 28 hours after the beginning of the study. The following concentrations were tested:

	without S9-mix (µg/ml)	with S9-mix (µg/ml)
7 hours	100	150
18 hours	15, 100, 150	15, 150, 200
28 hours	150	150

In the presence of S9-mix, 150 µg 3-nitro-1,2-dimethylbenzene/ml induced structural chromosome aberrations, and was thus mutagenic in this test system (CCR, 1988 b).

7.6.2 In vivo

No information available.

7.7   Carcinogenicity

No information available.

7.8   Reproductive toxicity

No information available.

7.9   Effects on the immune system

No information available.

7.10   Neurotoxicity

No information available.

7.11   Other effects

No information available.

8.   Experience in humans

No information available.

9.   Threshold limit values

No information available.

References

CCR (Cytotest Cell Research GmbH & Co KG)
Salmonella typhimurium and *Escherichia coli* reverse mutation assay with 3-Nitro-1,2-dimethylbenzol
Unpublished report, CCR project 116212 (1988 a)
On behalf of BG Chemie

CCR (Cytotest Cell Research GmbH & Co KG)
Chromosome aberration assay in Chinese hamster V79 cells in vitro with 3-Nitro-1,2-dimethylbenzol
Unpublished report, CCR project 116223 (1988 b)
On behalf of BG Chemie

Hoechst AG, Pharma Forschung Toxikologie und Pathologie
Nitro-o-xylol vic. - Prüfung der akuten oralen Toxizität an der
männlichen und weiblichen Wistar-Ratte
Unpublished report no. 87.0213 (1987 a)

Hoechst AG, Pharma Forschung Toxikologie und Pathologie
Nitro-o-xylol vic. - Prüfung auf Hautreizung am Kaninchen
Unpublished report no. 87.0108 (1987 b)

Hoechst AG, Pharma Forschung Toxikologie und Pathologie
Nitro-o-xylol vic. - Prüfung auf Augenreizung am Kaninchen
Unpublished report no. 87.0133 (1987 c)

Kawai, A., Goto, S., Matsumoto, Y., Matsushita, H.
Mutagenicity of aliphatic and aromatic nitro compounds
Jpn. J. Ind. Health (Sangyo Igaku), 29, 34–54 (1987)

Nohmi, T., Yoshikawa, K., Nakadate, M., Miyata, R., Ishidate, M., jr.
Mutations in *Salmonella typhimurium* and inactivation of *Bacillus
subtilis* transforming DNA induced by phenylhydroxylamine derivatives
Mutat. Res., 136, 159–168 (1984)

Thiem, K.W., Sewekow, B., Kiel, W., Handschuh, V., Freese, H.,
Schimpf, R., Vagt, H., Bunge, W.
Nitroverbindungen, aromatische
In: Ullmanns Encyklopädie der technischen Chemie
4th ed., vol. 17, p. 383–416
Verlag Chemie, Weinheim (1979)

TNO Nutrition and Food Research, Zeist, Netherlands
Short-term oral toxicity studies with 3-Nitro-1,2-dimethylbenzol in
rats: - range-finding (5-day) study, - sub-acute (28-day) study with a
14-day recovery study
Unpublished report no. V88.465 (1991)
On behalf of BG Chemie

2-Nitro-1,4-dimethylbenzene

1. Summary and assessment

In two in vitro mutagenicity studies in *Salmonella typhimurium* that were reported only in general terms, 2-nitro-1,4-dimethylbenzene has been shown to have mutagenic activity in strain TA 100 only in the presence of metabolic activation. As no further toxicological data are available, toxicological evaluation of 2-nitro-1,4-dimethylbenzene is not possible.

2. Name of substance

2.1	Usual name	2-Nitro-1,4-dimethylbenzene
2.2	IUPAC-name	2-Nitro-1,4-dimethylbenzene
2.3	CAS-No.	89–58–7
2.4	EINECS-No.	201–920–7

3. Synonyms, common and trade names

1,4-Dimethyl-2-nitrobenzene
2-Nitro-1,4-dimethylbenzol
2,5-Dimethylnitrobenzene
2-Nitro-p-xylene
Nitro-p-xylene

4.   Structural and molecular formulae

4.1   Structural formula

$$\text{C}_6\text{H}_3(\text{CH}_3)_2(\text{NO}_2)$$

4.2   Molecular formula $C_8H_9NO_2$

5.   Physical and chemical properties

5.1   Molecular mass, g/mol   151.16

5.2   Melting point, °C   -25 (Thiem et al., 1979)

5.3   Boiling point, °C   240 (at 1000 hPa)
                                      (Thiem et al., 1979)

5.4   Vapour pressure, hPa   No information available

5.5   Density, g/cm^3 1.132 (at 15 °C)
                                      (Thiem et al., 1979)

5.6   Solubility in water   Insoluble (Thiem et al., 1979)

5.7   Solubility in organic solvents   Dissolves readily in most
                                      organic solvents
                                      (Thiem et al., 1979)

5.8 Solubility in fat   No information available

5.9 pH-value   –

5.10 Conversion factor   1 ml/m^3 (ppm) $\triangleq$ 6.17 mg/m^3
                                      1 mg/m^3 $\triangleq$ 0.16 ml/m^3 (ppm)
                                      (at 1013 hPa and 25 °C)

6.   Uses

In the manufacture of p-xylidine, from which dyestuffs and pharmaceuticals are produced (Thiem et al., 1979)

7.   Experimental results

7.1   Toxicokinetics and metabolism

No information available.

7.2   Acute and subacute toxicity

No information available.

7.3   Skin and mucous membrane effects

No information available.

7.4   Sensitization

No information available.

7.5   Subchronic and chronic toxicity

No information available.

7.6   Genotoxicity

7.6.1 In vitro

2-Nitro-1,4-dimethylbenzene was studied in the Salmonella/micro-some test in strains TA 98 and TA 100 with and without metabolic activation (S9-mix from Kanechlor KC-400-induced rat liver). A mutagenic effect was seen in strain TA 100 after the addition of S9-mix (no further details; Nohmi et al., 1984).

In a further study in strains TA 100 and TA 98 with and without metabolic activation (from Kanechlor 500-induced rat liver), an increase in the revertant count (2500/mg 2-nitro-1,4-dimethylben-zene) was also seen in strain TA 100 with S9-mix (no further details; Kawai et al., 1987).

7.6.2 In vivo

No information available.

7.7   Carcinogenicity

No information available.

7.8   Reproductive toxicity

No information available.

7.9   Effects on the immune system

No information available.

7.10 Neurotoxicity

No information available.

7.11 Other effects

No information available.

8.   Experience in humans

No information available.

9.   Threshold limit values

No information available.

References

Kawai, A., Goto, S., Matsumoto, Y., Matsushita, H.
Mutagenicity of aliphatic and aromatic nitro compounds
Jpn. J. Ind. Health (Sangyo Igaku), 29, 34–54 (1987)

Nohmi, T., Yoshikawa, K., Nakadate, M., Miyata, R., Ishidate, M., jr.
Mutations in *Salmonella typhimurium* and inactivation of *Bacillus subtilis* transforming DNA induced by phenylhydroxylamine derivatives
Mutat. Res., 136, 159–168 (1984)

Thiem, K.W., Sewekow, B., Kiel, W., Handschuh, V., Freese, H., Schimpf, R., Vagt, H., Bunge, W.
Nitroverbindungen, aromatische
In: Ullmanns Encyklopädie der technischen Chemie
4th ed., vol. 17, p. 383–416
Verlag Chemie, Weinheim (1979)

4-Nitro-1,3-dimethylbenzene

1. Summary and assessment

4-Nitro-1,3-dimethylbenzene is of low to moderate acute toxicity (LD_{50} rat oral, males 2240 mg/kg body weight, females 1690 mg/kg body weight).

Repeated administration of 4-nitro-1,3-dimethylbenzene in the diet at 0, 200, 1000 or 5000 ppm for 5 days or at 0, 100, 600 or 3000 ppm for 28 days (equivalent to about 0, 20, 100 and 500 mg/kg body weight/day and 0, 10, 60 and 300 mg/kg body weight/day respectively) caused reduced body weight gain and feed intake and an increase in the relative liver weight in the preliminary study at 5000 ppm. In the 28-day study, effects observed at 3000 ppm included a decrease in body weight gain and feed intake, adverse effects on the red blood count (decrease in erythrocytes and in haemoglobin) and reversible functional disturbances of the liver (decrease in total protein and albumin, increase in absolute and relative liver weights). There were no histopathological changes in the liver, adrenal glands, kidneys, heart or spleen. The no toxic effect level is reported as 600 ppm (ca. 60 mg/kg body weight/day).

4-Nitro-1,3-dimethylbenzene is not irritating to rabbit skin.

Transient irritation of the conjunctiva is observed one hour after instillation into the rabbit eye. The substance has been evaluated as not irritating in accordance with EC classification criteria.

In the Salmonella/microsome test, 4-nitro-1,3-dimethylbenzene induces mutations in strain TA 100 with and without S9-mix and in strain TA 1535 without S9-mix, but it does not induce mutations in Chinese hamster V79 cells in the HPRT test either with or without S9-mix. The substance is mutagenic in the chromosome aberration test in Chinese hamster V79 cells in vitro, but not in the chromosome aberration test in vivo (bone marrow cells from the Chinese hamster).

2.   Name of substance

2.1   Usual name   4-Nitro-1,3-dimethylbenzene

2.2   IUPAC-name   4-Nitro-1,3-dimethylbenzene

2.3   CAS-No.   89–87–2

2.4   EINECS-No.   201–947–4

3.   Synonyms, common and trade names

1,3-Dimethyl-4-nitrobenzene
2,4-Dimethyl-1-nitrobenzene
4-Nitro-1,3-dimethylbenzol
4-Nitro-m-xylene
1-Nitro-2,4-dimethylbenzene
Nitro-m-xylene, asym.
Nitro-m-xylene (413)

4.   Structural and molecular formulae

4.1   Structural formula

4.2   Molecular formula $C_8H_9NO_2$

5.   Physical and chemical properties

5.1   Molecular mass, g/mol   151.16

5.2   Melting point, °C   9 (Thiem et al., 1979)

5.3   Boiling point, °C   244 (at 1013 hPa)
                                 (Thiem et al., 1979)

5.4	Vapour pressure, hPa	1 (at 60.3 °C) 13 (at 109.8C) (Thiem et al., 1979)
5.5	Density, g/cm^3	1.126 (at 20 °C) (Thiem et al., 1979).
5.6	Solubility in water	Insoluble (Thiem et al., 1979)
5.7	Solubility in organic solvents	Dissolves readily in most organic solvents (Thiem et al., 1979)
5.8	Solubility in fat	No information available
5.9	pH-value	–
5.10	Conversion factor	1 ml/m^3 (ppm) $\triangleq$ 6.17 mg/m^3 1 mg/m^3 $\triangleq$ 0.16 ml/m^3 (ppm) (at 1013 hPa and 25 °C)

6.   Uses

In the manufacture of 2,4-xylidine, from which dyestuffs, insecticides, herbicides and veterinary medicines are produced (Thiem et al., 1979).

7.   Experimental results

7.1   Toxicokinetics and metabolism

No information available.

7.2   Acute and subacute toxicity

The acute oral toxicity of 4-nitro-1,3-dimethylbenzene was studied in Wistar rats (5 males and 5 females/dose). LD_{50} values of 2240 (1840 to 2730) mg/kg for males and 1690 (1270 to 2030) mg/kg for females were calculated. 4-Nitro-1,3-dimethylbenzene was thus shown to be of low toxicity. Signs of toxicity observed included disturbances in movement, impairment of breathing and reflexes and stupefaction, all of which were reversible after day 4 of the study. In the animals that died, the liver and kidneys were discoloured and the lungs congested. There were no visible macroscopic effects in the rats killed at the end of the 14-day observation period (Hoechst, 1986 a).

In a range-finding study carried out prior to a 28-day feeding study, groups of 5 male and 5 female Wistar rats (3 to 4 weeks old) were given 4-nitro-1,3-dimethylbenzene in the diet at 0, 200, 1000 or 5000 ppm for 5 days (99.4% pure; equivalent to 0, 20, 100 and 500 mg/kg body weight/day, respectively). The homogeneous distribution of the test substance and its stability in the feed complied with requirements. Body weight was reduced in males and feed intake was reduced in both sexes at the highest dose level at the end of the 5-day study. The haemoglobin concentration was significantly higher in males at the 1000 and 5000 ppm doses than in the controls. The relative liver weight was increased in males at the top dose level. No effects were found at post-mortem (TNO, 1991).

Dietary concentrations of 0, 100, 600 and 3000 ppm 4-nitro-1,3-dimethylbenzene (99.4% pure; equivalent to 0, 10, 60 and 300 mg/kg body weight/day, respectively) were chosen for the subsequent 28-day feeding study, which was carried out in accordance with OECD guideline no. 407. The homogeneous distribution of the test substance and its stability in the feed complied with requirements. The dose groups each consisted of 5 male and 5 female rats; an additional 5 males and 5 females were used in the 0 and 3000 ppm groups and kept for a 14-day observation period. Body weight was significantly reduced in both sexes at the high dose level. The difference from the controls was also apparent in the observation period, but was only significant in the males. Feed intake was slightly reduced in the males in the high dose group only. Erythrocytes and haemoglobin were reduced compared with the control values in the high-dose males. The thrombocyte count was increased. No corresponding effects were found in the females. There were no treatment-related effects on the leukocytes or on the differential blood count. Total protein and albumin were significantly reduced compared with the controls in both sexes at the high-dose level at the end of treatment, an effect that was only reversible in the females during the observation period. The activities of alanine- and aspartate aminotransferase (ALT and AST) were relatively low in the high-dose males at the end of the study. Urinalysis revealed no differences between the control and treated animals. The absolute and relative liver weights were significantly increased in both sexes at the high dose level at the end of the 28-day feeding period, but only remained elevated in the males at the end of the observation period. At the end of the treatment period, the absolute and relative adrenal gland weights were increased in the high-dose males, while in the females there was a significant increase in the absolute and relative spleen weights in the 3000 ppm group. No effects that could be regarded as treatment-related were evident on histological examination of the liver, kidneys, heart, adrenal glands and spleen at the end of the feeding period. The no toxic effect level determined in this 28-day study was 600 ppm (equivalent to ca. 60 mg/kg body weight/day; TNO, 1991).

7.3   Skin and mucous membrane effects

The acute skin irritancy of > 99% pure 4-nitro-1,3-dimethylbenzene to the clipped dorsal skin of New Zealand albino rabbits was tested in accordance with OECD guideline no. 404 (0.5 ml/rabbit, 4-hour exposure, neat substance, semi-occlusive cover). No irritant effects were found during the 72-hour observation period (Hoechst, 1986 b).

The acute eye irritancy of 4-nitro-1,3-dimethylbenzene (> 99% pure) was tested in New Zealand albino rabbits in accordance with OECD guideline no. 405 (0.1 ml neat substance/eye). One hour after instillation, the conjunctiva showed marked congestion of the blood vessels and was a diffuse, crimson colour with a clear discharge. All of the observed irritant effects were reversible from 2 to 4 days after instillation. The substance was evaluated as not irritating in accordance with the guidelines contained in 83/467/EEC (Hoechst, 1986 c).

7.4 Sensitization

No information available.

7.5   Subchronic and chronic toxicity

No information available.

7.6   Genotoxicity

7.6.1 In vitro

4-Nitro-1,3-dimethylbenzene (99.4% pure) was studied in the plate incorporation test with and without metabolic activation (S9-mix from Aroclor 1254-induced rat liver) in the *Salmonella typhimurium* strains TA 1535, TA 1537, TA 1538, TA 98 and TA 100 and in *Escherichia coli* WP2. Concentrations of from 3.3 to 5000 µg/plate were used. Cytotoxic effects were observed from 1000 µg/plate both with and without S9-mix. A dose-dependent increase in the revertant count (by up to a factor of 3) was seen up to a concentration of 1000 µg/plate in strain TA 100 with and without S9-mix and in strain TA 1535 without metabolic activation. 4-Nitro-1,3-dimethylbenzene was thus mutagenic in this study (CCR, 1988 a).

4-Nitro-1,3-dimethylbenzene was positive in the Salmonella/microsome test in strain TA 100 with S9-mix (from Kanechlor KC-400-induced rat liver). No increase in the mutant count was seen without S9-mix. The compound was negative in strain TA 98 with and without S9-mix (no further details; Nohmi et al., 1984).

In a further study in strains TA 100 and TA 98, 4-nitro-1,3-dimethylbenzene caused a dose-dependent increase in revertants only in the presence of S9-mix (from Kanechlor 500-induced rat liver). The increase did not, however, exceed a factor of 2 (no further details; Kawai et al., 1987).

In the Salmonella/microsome test in strains TA 98, TA 100, TA 1535, TA 1537 and TA 1538, pure 4-nitro-1,3-dimethylbenzene (> 99.9%) caused a dose-dependent increase in the revertant count in all trials in strain TA 100 only, both with and without S9-mix (from Aroclor 1254-induced rat liver; concentrations 4 to 5000 µg/plate). However, the increase only exceeded a factor of 2 in 3 of the 6 trials. The test substance was evaluated as mutagenic by the authors (Hoechst, 1990).

In a chromosome aberration test in vitro in Chinese hamster V79 cells carried out at concentrations of from 5 to 100 µg/ml with and without S9-mix (from Aroclor 1254-induced rat liver), 4-nitro-1,3-dimethylbenzene (99.4% pure) induced structural aberrations at a concentration of 100 µg/ml with S9-mix (CCR, 1988 b).

4-Nitro-1,3-dimethylbenzene (99.98% pure) was also tested in two independently-performed HPRT tests to detect gene mutations in Chinese hamster V79 cells with and without metabolic activation (S9-mix from Aroclor 1254-induced rat liver) at concentrations of 10, 30, 60, 80 and 110 µg/ml. The test substance showed no mutagenic activity in these studies (CCR, 1992).

7.6.2 In vivo

Groups of 5 male and 5 female Chinese hamsters were given a single oral dose of 4000 mg 4-nitro-1,3-dimethylbenzene (99.98% pure)/kg body weight in polyethylene glycol 400 (maximum tolerated dose). The bone marrow was examined for chromosome aberrations after 6, 24 and 48 hours. The evaluation of 50 metaphases/animal showed that 4-nitro-1,3-dimethylbenzene did not induce chromosome aberrations (CCR, 1991).

7.7   Carcinogenicity

No information available.

7.8   Reproductive toxicity

No information available.

7.9   Effects on the immune system

No information available.

7.10 Neurotoxicity

No information available.

7.11 Other effects

No information available.

8.   Experience in humans

No information available.

9.   Threshold limit values

No information available.

References

CCR (Cytotest Cell Research GmbH & Co KG)
Salmonella typhimurium and *Escherichia coli* reverse mutation assay with 4-Nitro-1,3-dimethylbenzol
Unpublished report, CCR project 115918 (1988 a)
On behalf of BG Chemie

CCR (Cytotest Cell Research GmbH & Co KG)
Chromosome aberration assay in Chinese hamster V79 cells in vitro with 4-Nitro-1,3-dimethylbenzol
Unpublished report, CCR project 115920 (1988 b)
On behalf of BG Chemie

CCR (Cytotest Cell Research GmbH & Co KG)
Chromosome aberration assay in bone marrow cells of the Chinese hamster with Nitro-m-xylol
Unpublished report, CCR project 187020 (1991)
On behalf of BG Chemie

CCR (Cytotest Cell Research GmbH & Co KG)
Gene mutation assay in Chinese hamster V79 cells in vitro with Nitro-m-xylol
Unpublished report, CCR project 187018 (1992)
On behalf of BG Chemie

Hoechst AG, Pharma Forschung Toxikologie
Nitro-m-xylol asym. - Prüfung der akuten oralen Toxizität an der

männlichen und weiblichen Wistar-Ratte
Unpublished report no. 86.0474 (1986 a)

Hoechst AG, Pharma Forschung Toxikologie
Nitro-m-xylol asym. - Prüfung auf Hautreizung am Kaninchen
Unpublished report no. 86.0384 (1986 b)

Hoechst AG, Pharma Forschung Toxikologie
Nitro-m-xylol asym. - Prüfung auf Augenreizung am Kaninchen
Unpublished report no. 86.0385 (1986 c)

Hoechst AG, Pharma Research Toxicology and Pathology
Nitro-m-xylol asym. - Study of the mutagenic potential in strains of
Salmonella typhimurium (Ames test)
Unpublished report no. 89.1210 (1990)

Kawai, A., Goto, S., Matsumoto, Y., Matsushita, H.
Mutagenicity of aliphatic and aromatic nitro compounds
Jpn. J. Ind. Health (Sangyo Igaku), 29, 34–54 (1987)

Nohmi, T., Yoshikawa, K., Nakadate, M., Miyata, R., Ishidate, M., jr.
Mutations in *Salmonella typhimurium* and inactivation of *Bacillus sub-
tilis* transforming DNA induced by phenylhydroxylamine derivatives
Mutat. Res., 136, 159–168 (1984)

Thiem, K.W., Sewekow, B., Kiel, W., Handschuh, V., Freese, H.,
Schimpf, R., Vagt, H., Bunge, W.
Nitroverbindungen, aromatische
In: Ullmanns Encyklopädie der technischen Chemie
4th ed., vol. 17, p. 383–416
Verlag Chemie, Weinheim (1979)

TNO Nutrition and Food Research, Zeist, Netherlands
Short-term oral toxicity study with 4-Nitro-1,3-dimethylbenzol in
rats: - range-finding (5-day) study, - sub-acute (28-day) study with a
(14-day) recovery study
Unpublished report no. V88.462 (1991)
On behalf of BG Chemie

Chloroacetic acid methyl ester

This Toxicological Evaluation replaces a previously published version in volume 3.

1.   Summary and assessment

Chloroacetic acid methyl ester is toxic on acute oral or dermal exposure (LD_{50} rat oral, 107 and 140 mg/kg body weight; rat dermal 136.6 mg/kg and ca. 470 mg/kg body weight; rabbit dermal 318 mg/kg body weight). Both non-specific signs of toxicity and clear irritant effects are observed. Comparable symptoms also occur after intraperitoneal injection (LD_{50} mouse i.p. 200 to 460 mg/kg body weight) and vapour inhalation. Marked irritation of the respiratory tract and eyes is also observed on inhalation exposure (LC_{50} rat > 945 < 1418 mg/m^3).

After repeated inhalation exposure to chloroacetic acid methyl ester for 28 days at concentrations of 10, 33 or 100 ppm (equivalent to 44, 146 and 443 mg/m^3, respectively), adverse effects on the breathing (irregular breathing) and on movement (uncoordinated gait) as well as irritant effects (closing together of the eyelids, sneezing, increased frequency of cleaning) and an increase in relative lung weight occurred in rats at the highest concentration. These effects, though less pronounced, were also observed in males in the 33 ppm group. In contrast, no such effects were detected in the 10 ppm group. Body weight gain was markedly impaired in both sexes at the highest concentration. Thus in this study, the no effect level is 10 ppm in male rats and 33 ppm in female rats.

Chloroacetic acid methyl ester is corrosive to rabbit skin and is absorbed through the skin (0.5 ml undiluted substance per rabbit leads to death). The substance is severely irritating to the rabbit eye.

In guinea-pigs, chloroacetic acid methyl ester induces sensitization. Cross-sensitization with chloroacetic acid ethyl ester has been observed in this species.

Chloroacetic acid methyl ester is not mutagenic in the Salmonella/microsome test in five strains of *Salmonella typhimurium* or in *Escherichia coli* either with or without metabolic activation. No induction of micronuclei has been observed in the micronucleus test in the mouse.

No increase in the incidence of lung tumours has been established after intraperitoneal administration to strain A mice.

In man, delayed occurrence of irritation of the conjunctiva is observed after exposure to the vapour.

2.   Name of substance

2.1	Usual name	Chloroacetic acid methyl ester
2.2	IUPAC-name	Chloroacetic acid methyl ester
2.3	CAS-No.	96–34–4
2.4	EINECS-No.	202–501–1

3.   Synonyms, common and trade names

Chloressigsäuremethylester
Acetic acid, chloro-, methyl ester
Methyl -α-chloroacetate
Methylchloroacetate
Methylmonochloroacetate
Methyl monochloroacetate
Monochloroacetic acid methyl ester

4.   Structural and molecular formulae

4.1   Structural formula $Cl\text{–}CH_2\text{–}COOCH_3$

4.2   Molecular formula $C_3H_5ClO_2$

5.   Physical and chemical properties

5.1   Molecular mass, g/mol   108.52

5.2   Melting point, °C   –32.7 (Koenig et al., 1986)

5.3   Boiling point, °C   129.8 (Weast, 1981/82)
130–131 (Koenig et al., 1986)

5.4   Vapour pressure, hPa   6.9 (at 20 °C) (Hoechst, 1990)

5.5   Density, g/cm^3 1.236 (at 20 °C)
(Koenig et al., 1986)

5.6   Solubility in water   2.8 g/l (at 20 °C) (Hoechst, 1990)

5.7   Solubility in organic solvents   Dissolves well in alcohol
and ether (Koenig et al., 1986)

5.8   Solubility in fat   No information available

5.9   pH-value   –

5.10   Conversion factor   1 ml/m^3 (ppm) $\triangleq$ 4.43 mg/m^3
1 mg/m^3 $\triangleq$ 0.23 ml/m^3 (ppm)
(at 1013 hPa and 25 °C)

6.   Uses

Intermediate in the production of pharmaceuticals (antiphlogistics, vitamin B_2, barbiturates), pesticides (dimethoate), malonic acid ester derivatives, fragrances, heterocyclenes; condensation with urea derivatives (Hoechst, 1991).

7.   Experimental results

7.1   Toxicokinetics and metabolism

No information available.

7.2   Acute and subacute toxicity

An oral LD_{50} value of 107 mg/kg body weight was reported after administration of a single dose of chloroacetic acid methyl ester as a 0.5% solution in sesame oil to female rats (weight 202 to 244 g). The observation period was 14 days. The rats died between 50 minutes and 24 hours after administration and showed the following clinical signs of toxicity: passivity, disturbances of balance, squatting behaviour, prostration, ruffled fur, diarrhoea and spasmodic breathing. Similar but less pronounced signs of toxicity were seen in the surviving rats, which showed no signs of toxicity 24 hours after administration. Post-mortems on the animals that died revealed the following macroscopic effects: the liver, spleen and kidneys were coloured dark brown and severely vascularized and the ovaries were reddened. No conspicuous effects were found at post-mortem in the surviving rats (Hoechst, 1979 a).

A further LD_{50} value of 140 mg/kg body weight was determined after a 14-day observation period in male and female rats following oral administration of chloroacetic acid methyl ester in olive oil. Clinical signs of toxicity included dyspnoea, apathy, abnormal positions, reeling, unkempt coats and diarrhoea. Dilation of the heart, congestive hyperaemia, bloody, corrosive gastritis and diarrhoeal gut contents were observed macroscopically in the rats that died. The organs of the animals that were killed at the end of the observation period showed no abnormal effects (BASF, 1981).

An oral LD_{50} value of 240 mg/kg body weight has been reported for the mouse (no further details; Izmerov, 1982).

On acute dermal administration of chloroacetic acid methyl ester the LD_{50} in the rat (sex not specified) was ca. 470 mg/kg body weight (BASF, 1981), that in the female Wistar rat was 136.6 mg/kg body weight (Hoechst, 1979 b) and that in female rabbits was 318 mg/kg body weight (Hoechst, 1979 d). On dermal application, skin erythema and oedema formation with subsequent necrosis and partial sloughing of the skin were seen. Autopsy of the animals that died revealed reddening of the small intestine with expansion of the blood vessels and dark-brown coloured and severely vascularized livers. Post-mortems of the surviving animals revealed no macroscopically visible effects (Hoechst, 1979 b, d).

The average lethal dose after a single intraperitoneal injection was between 200 and 460 mg/kg body weight in mice of both sexes (BASF, 1981).

In an inhalation hazard test (exposure times of 3, 10, 30 and 50 minutes) the rats died after exposure to an atmosphere enriched with chloroacetic acid methyl ester vapour at 20 °C for 10 minutes or more. In addition to narcotic effects, symptoms of irritation of the respiratory system and eyes occurred. Autopsies of the rats that died revealed acute dilation of the heart, acute congestive hyperaemia, severely oedematous lungs, multiple patch-like areas of congestion and multiple, severe acute swelling of the lungs. There were no abnormal effects on the organs of the animals killed at the end of the study (BASF, 1981).

In a further inhalation hazard test (exposure times 3, 10 and 30 minutes) with groups of 5 male and 5 female Wistar rats in accordance with OECD guideline no. 403, the rats died after inhalation of an atmosphere enriched with chloroacetic acid methyl ester vapour at 20 °C for periods of 3 minutes or greater. In addition to non-specific signs of toxicity, the animals showed impairment of breathing and coordination. In addition, partly-closed eyes, salivation, nasal secretion, corneal clouding and a weakening of the reflexes were observed. Macroscopically, effects on the lungs such as red to dark red staining, sometimes with dark foci, and the appearance of foamy fluid on cutting into the lungs were found in the rats that died and occasionally in those killed at the end of the study (Hoechst, 1987).

The LC_{50} in Wistar rats of both sexes (weight 205 to 212 g) after acute inhalation exposure for 4 hours was between 210 and 315 ppm (equivalent to 945 and 1418 mg/m^3). Signs of toxicity observed in addition to non-specific symptoms included impairment of breathing, coordination and reflexes. Furthermore, partly-closed eyes, sneezing, increased cleaning behaviour, red encrusted noses and clouding of the cornea were seen. None of the surviving animals showed clinical signs of toxicity 14 days after inhalation. The rats that died showed macroscopic effects on the lungs and gastrointestinal tract. At the end of the study, effects on the lungs such as enlargement and dirty-red, light beige or orange staining (sometimes with red foci) were found (Hoechst, 1988 a).

Groups of 4 to 5 female rats (strain not specified) were acutely exposed to chloroacetic acid methyl ester (purity > 95%) by inhalation in a 160 l exposure chamber (air flow 15 to 25 l/min). Concentrations, inhalation times and mortality are shown in Table 1.

Table 1. Mortality after acute exposure to various concentrations of chloroacetic acid methyl ester

Concentration (ppm)		Exposure time (hours)	Number of dead/exposed animal
calculated	measured		
1000	920	4.0	5/5
1000	–	2.0	4/5
1000	–	1.0	2/4
500	400	4.0	5/5
500	–	2.0	0/4
250	–	7.0	1/5
250	–	4.0	0/4
100	81	7.0	0/4

– not measured

Deaths occurred a few days after exposure, and the authors considered that these were possibly due to pneumonia. Chloroacetic acid methyl ester was severely irritating to the airways at concentrations of 250 to 1000 ppm. At a concentration of 100 ppm, only a slight irritant effect was seen (Torkelson et al., 1971).

Under the same experimental protocol, groups of rabbits (number, strain and sex not given) were exposed to chloroacetic acid methyl ester for 7 or 4 hours at 100 and at 50 ppm. The animals exhibited irritation of the conjunctiva and cornea at 100 ppm, but not at 50 ppm. The observed irritation was reversible; staining of the cornea with fluorescein 1 week after exposure showed no corneal damage. No significant weight loss or deaths were observed (Torkelson et al., 1971).

In a subacute inhalation study, groups of 10 male and 10 female Wistar rats (ca. 45 days old, weight 144 to 161 g) were exposed 20 times in total to chloroformic acid methyl ester at concentrations of 10, 33 or 100 ppm (equivalent to 44, 146 and 443 mg/m^3, respectively) for 6 hours/day, 5 days a week for 4 weeks. A control group was simultaneously exposed to clean air under identical conditions. 5 males and 5 females from each group were examined macroscopically and microscopically at the end of the study, the remaining rats after a 14-day recovery period. Results showed eye irritation in the 10 ppm group on the first day of the study only. The animals in the 33 ppm group exhibited eye irritation, sneezing and increased frequency of cleaning during exposure. These symptoms were not observed prior to exposure or on the exposure-free days. None of the experimental animals in the 10 and 33 ppm groups showed clinical symptoms during the recovery period. The animals in the 100 ppm group showed eye irritation, sneezing, squatting behaviour, increased frequency of cleaning, irregular breathing, uncoordinated movement, retracted flanks, reduced activity and ruffled fur both during and after exposure. The symptoms were occasionally still present on the following day before exposure, and could still be observed in this dose group during the recovery period, up to day 35 (1 week after the last exposure). No treatment-related effects were seen on the central nervous system, eyes, teeth or mucous membranes of the mouth. Animals in the 100 ppm group, particularly the males, showed a clear impairment of body weight gain, although this was partially reversible in the males and completely reversible in the females after the recovery period. Weight gain was also slightly retarded in the males of the 33 ppm group. Animals in the 100 ppm group had reduced food intake and increased drinking water requirements. Haematological and clinical-chemistry investigations gave no indications of treatment-related effects on the experimental animals. Urinalysis also revealed no effects. Exposure to 100 ppm chloroacetic acid methyl ester led to a clear increase in relative lung weight. This effect was only partly reversible 14 days after the end of exposure. Autopsy of the animals revealed no conspicuous treatment-related macroscopic changes, and microscopic examination of the organ tissues gave no indication of treatment-related effects. The authors gave a no effect level for the systemic toxic effects of 33 ppm for the females and 10 ppm for the males (Hoechst, 1988 b).

7.3   Skin and mucous membrane effects

A patch test for skin irritation carried out in accordance with FDA guidelines (0.5 ml undiluted product) caused the death of the treated rabbits (Albino-Himalaya; weight 1.5 to 2.5 kg) between 2 and 8 hours after application. In a modified study, the application of 125 mg/kg body weight as the neat substance (maximum tolerated dose; occlusive, 24 hours) produced severe damage to the skin in the form of redness, swelling, scab formation, cracking, hardening, scaling and corrosion. The substance was a severe skin irritant and corrosive under the study conditions described (Hoechst, 1979 c).

In a further study protocol, chloroacetic acid methyl ester was applied to the skin of groups of 2 to 4 rabbits for 3 minutes or for 1 or 4 hours (no further details). After exposure for 3 minutes, slight erythema occurred. After exposure for 1 or 4 hours, mild to very severe erythema and mild to severe oedema were observed. Necrosis was reported from the second day after application. In a study of skin irritation, conducted in accordance with FDA guidelines, groups of 3 male and 3 female rabbits (White Viennese) had 0.5 ml of the test substance applied to the intact or scarified skin. Severe to very severe erythema and oedema as well as necrosis were observed on both the intact and scarified skin. The primary irritation value was between 6.6 and 6.8. Based on these results the substance was evaluated as corrosive (BASF, 1981).

6 rabbits each had 0.1 ml of the neat test substance instilled into the conjunctival sac of one eye in accordance with FDA guidelines. After 72 hours the highest irritation index of 64 was determined. Based on these results, the substance was evaluated as severely irritating to the mucous membranes (Hoechst, 1979 c).

Groups of 3 male and 3 female rabbits (White Viennese) had 0.1 ml of "crude" chloroacetic acid methyl ester instilled into the conjunctival sac of one eye in accordance with FDA guidelines.

After 24 hours, distinct clouding of the cornea, ciliary injection of the iris, marked reddening and swelling of the conjunctiva and marked to severe increases in secretion were observed in all animals. After 15 days, slight to distinct clouding of the cornea, slightly intensified to marked reddening of the conjunctiva and, in 2 rabbits, slightly increased secretion were still present. Marginal vascularization of the cornea and scar formation were evident in all rabbits. Based on a primary irritation index of 57, the substance was evaluated as severely irritating (BASF, 1981).

7.4   Sensitization

The ability of chloroacetic acid methyl ester to induce sensitization was studied in the Magnusson-Kligman maximization test in 20 female guinea-pigs (Dunkin-Hartley). Intradermal induction was carried out at a concentration of 0.05% (w/w) in liquid paraffin, dermal induction at a concentration of 2% (w/w). After challenges with concentrations of 1, 0.1 and 0.01% (w/w), 13 of the 20 treated animals showed positive reactions with scaling or scab formation in response to a concentration of 1%. In contrast, none of the control animals gave a positive reaction. The substance was thus shown to induce sensitization (NOTOX, 1987).

To study possible cross-sensitization with carbonic acid ethyl ester, the ability of chloroacetic acid ethyl ester to induce skin sensitization was studied in a Magnusson-Kligman maximization test in 20 female Dunkin-Hartley guinea-pigs. After the induction treatment (intradermal 5% (v/v) in 50% ethanol; dermal 2.5% (v/v) in 50% ethanol) and 2 challenge treatments (0.025 to 5% v/v) in 50% ethanol; sensitization rate 79%), 4 guinea-pigs from this study were treated on day 49, with chloroacetic acid methyl ester at concentrations of 0.05, 0.5 and 5% (v/v) in 50% ethanol, in order to study possible cross-reactivity. All 4 guinea-pigs showed a positive reaction (NOTOX, 1986; Braun and van der Walle, 1987).

7.5 Subchronic and chronic toxicity

No information available.

7.6 Genotoxicity

7.6.1 In vitro

Chloroacetic acid methyl ester (purity 99.4%) was tested for mutagenic activity in the Salmonella/microsome test in the *Salmonella typhimurium* strains TA 98, TA 100, TA 1535, TA 1537 and TA 1538 and in *Escherichia coli* WP2uvrA, with and without metabolic activation (S9-mix from liver homogenates from Aroclor 1254 pre-treated rats). Concentrations of from 4 to 5000 µg/plate were used. The product was bacteriotoxic from 2500 µg/plate. The substance was not mutagenic at the concentrations tested, either in the presence or absence of a metabolic activation system (Hoechst, 1983).

Chloroacetic acid methyl ester was also not mutagenic in a further Salmonella/microsome test in the *Salmonella typhimurium* strain TA 100 with or without metabolic activation (no further details; Sato et al., 1985).

In a more recent Salmonella/microsome test, chloroacetic acid methyl ester (99.7% pure) was tested at doses of 4, 20, 100, 500, 2500, 5000 and 10,000 µg/plate in one experiment and 4, 20, 100, 500, 2500 and 5000 µg/plate in a second experiment in the *Salmonella typhimurium* strains TA 98, TA 100, TA 1535, TA 1537 and TA 1538 with and without metabolic activation (S9-mix from Aroclor 1254-induced male rat liver). The substance was cytotoxic at 2500 µg/plate and above. No significant increase in the revertant count was detected either with or without metabolic activation. Chloroacetic acid methyl ester was not mutagenic in this test system (Hoechst, 1993).

7.6.2 In vivo

A micronucleus test was carried out with chloroacetic acid methyl ester. The chemical was administered as a single dose by stomach tube to male and female NMRI-mice (7 weeks old, weight 21 to 33 g), at the maximum tolerated dose of 300 mg/kg body weight. Positive controls received Endoxan (50 mg/kg body weight). 5 animals of each sex were killed 24, 48 and 72 hours after administration and the bone marrow was examined. No induction of micronuclei was found. The ratio of polychromatic to normochromatic erythrocytes was not adversely affected by the treatment (Hoechst, 1988 c).

7.7 Carcinogenicity

Groups of 20 mice (strain A; 10 males and 10 females per group) were given chloroacetic acid methyl ester 3 times per week for 8 weeks by intraperitoneal injection at doses of 0.36, 0.18 and 0.09 mmol/kg body weight, equivalent to 39 mg/kg body weight (the maximum tolerated dose), 19.5 and 9.7 mg/kg body weight, respectively. The observation period was 16 weeks. With the exception of 1 mouse in the low-dose group, all of the mice survived. No increase in the incidence of lung tumours was detected (Theiss et al., 1979). This carcinogenicity test involves the induction of lung tumours, which have a high spontaneous incidence in the mouse strain used, and which occur more frequently on administration of a carcinogen. The value of this method is limited.

7.8 Reproductive toxicity

No information available.

7.9 Effects on the immune system

No information available.

7.10 Neurotoxicity

No information available.

7.11 Other effects

No information available.

8. Experience in humans

Industrial experience has shown that the chemical causes the delayed occurrence of corneal irritation in humans (no further details; Torkelson et al., 1971).

9.   Threshold limit values

No information available.

References

BASF AG, Gewerbehygiene und Toxikologie
Gewerbetoxikologische Grundprüfung
Unpublished report (1981)

Braun, C.L.J., van der Walle, H.B.
The ethylester of monochloroacetic acid
Contact Dermatitis, 16, 114–115 (1987)

Hoechst AG, Pharma Forschung Toxikologie
Akute orale Toxizität von Monochloressigsäuremethylester an weib-
lichen Ratten
Unpublished report no. 139/79 (1979 a)

Hoechst AG, Pharma Forschung Toxikologie
Akute dermale Toxizität von Monochloressigsäuremethylester an
weiblichen Ratten
Unpublished report no. 140/79 (1979 b)

Hoechst AG, Pharma Forschung Toxikologie
Haut- und Schleimhautverträglichkeit von Monochloressigsäure-
methylester an Kaninchen
Unpublished report no. 141/79 (1979 c)

Hoechst AG, Pharma Forschung Toxikologie
Akute dermale Toxizität von Monochloressigsäuremethylester an
Kaninchen
Unpublished report no. 142/79 (1979 d)

Hoechst AG, Pharma Forschung Toxikologie
Monochloressigsäuremethylester - Study of the mutagenic potential

in strains of *Salmonella typhimurium* (Ames test) and Escherichia coli
Unpublished report no. 83.0084 (1983)

Hoechst AG, Pharma Forschung Toxikologie
Monochloressigsäuremethylester - Inhalationstoxizität im Zeitsättigungstest an männlichen und weiblichen SPF-Wistar Ratten
Unpublished report no. 87.1414 (1987)

Hoechst AG, Pharma Forschung Toxikologie und Pathologie
Monochloressigsäuremethylester - Inhalation im strömenden Gemisch
an männlichen und weiblichen SPF-Wistar Ratten, 4 Stunden-LC_{50}
Unpublished report no. 88.0041 (1988 a)
On behalf of BG Chemie

Hoechst AG, Pharma Forschung Toxikologie und Pathologie
Chloressigsäuremethylester - Subakute Inhalation (20 Applikationen in 28 Tagen) an SPF-Wistar Ratten
Unpublished report no. 88.0233 (1988 b)
On behalf of BG Chemie

Hoechst AG, Pharma Forschung Toxikologie und Pathologie
Monochloressigsäuremethylester - Micronucleus test in male and
female NMRI mice after oral administration
Unpublished report no. 88.1329 (1988 c)

Hoechst AG
Safety data sheet Chloressigsäuremethylester (1990)

Hoechst AG
AIDA-Grunddatensatz Acetic acid, chloro-, methyl ester (1991)

Hoechst AG, Pharma Development Central Toxicology
Monochloressigsäuremethylester - Study of the mutagenic potential
in strains of *Salmonella typhimurium* (Ames test)
Unpublished report no. 92.0822 (1993)

Izmerov, N.F., Sanotsky, I.V., Sidorov, K.K.
Toxicometric parameters of industrial toxic chemicals under single

exposure (English translation from the Russian)
Centre of International Projects, GKNT, Moscow (1982)

Koenig, G., Lohmar, E., Rupprich, N.
Chloroacetic acids
In: Ullmann's encyclopedia of industrial chemistry
5th ed., vol. A6, p. 537–552
VCH-Verlagsgesellschaft, Weinheim (1986)

NOTOX C.V., The Netherlands
Assessment of the skin sensitization potential of monochloro acetic
acid ethyl ester and screening for cross-sensitivity to monochloro
acetic acid and monochloro acetic acid methyl ester in the guinea-
pig (Magnusson and Kligman Maximization test)
Unpublished report no. 0278/344 (1986)
On behalf of Medische Dienst ENKA, Arnhem, The Netherlands

NOTOX C.V., The Netherlands
Assessment of the skin sensitization potential of Azonal in the guin-
ea-pig (Magnusson and Kligman Maximization test)
Unpublished report no. 0603/758 (1987)
On behalf of Akzo N.V., Arnhem, The Netherlands

Sato, T., Mukaida, M., Ose, Y., Nagase, H., Ishikawa, T.
Mutagenicity of chlorinated products from soil humic substances
Sci. Total Environ., 46, 229–241 (1985)

Theiss, J.C., Shimkin, M.B., Poirier, L.A.
Induction of pulmonary adenomas in strain A mice by substituted
organohalides
Cancer Res., 39, 391–395 (1979)

Torkelson, T.R., Kary, C.D., Chenoweth, M.B., Larsen, E.R.
Single exposure of rats to the vapours of trace substances in
methoxyflurane
Toxicol. Appl. Pharmacol., 19, 1–9 (1971)

Weast R.C. (ed.)
CRC handbook of chemistry and physics
62nd ed., C-227
CRC Press, Boca Raton, Florida (1981/82)

Trichlorophenylsilane

Toxicological Evaluations are also available for the structurally related compounds dichlorodimethylsilane and trichloromethylsilane.

1.   Summary and assessment

The acute toxicity of trichlorophenylsilane is low in rats treated by stomach tube (oral LD_{50} 2390 mg/kg body weight), moderate in rabbits treated dermally (24-hour LD_{50} 0.89 ml (1166 mg)/kg body weight), and high in mice exposed by inhalation (2-hour LC_{50} 0.33 mg/l) or by intravenous injection (LD_{50} 100 mg/kg body weight). The mice exposed to low vapour concentrations (0.12–0.3 mg/l) showed signs of mucosal irritation (breathlessness, asphyxia, cyanosis) as well as agitated motor activity, convulsions, spreading of hind limbs), and those that died displayed various changes in the lungs, liver and brain. In an inhalation hazard test, 6 male albino rats survived exposure to the concentrated vapour for 8 hours.

Application of the liquid to the skin or eyes of rabbits produces corrosion, and the vapour is also irritating to the eyes of mice and rabbits.

2.   Name of substance

2.1	Usual name	Trichlorophenylsilane
2.2	IUPAC-name	Phenyltrichlorosilane
2.3	CAS-No.	98–13–5
2.4	EINECS-No.	202–640–8

3.   Synonyms, common and trade names

Phenylsilicontrichloride
Phenylsilicon trichloride
Silane, trichlorophenyl-
Silicon phenyl trichloride
Trichlorphenylsilan
Wacker Silan P

4.   Structural and molecular formulae

4.1   Structural formula

4.2   Molecular formula $C_6H_5Cl_3Si$

5.   Physical and chemical properties

5.1   Molecular mass, g/mol   211.6

5.2   Melting point, °C   - 40 (Wacker, 1991)

5.3   Boiling point, °C   201 (Wacker, 1991)

5.4   Vapour pressure, hPa   < 1 (at 20 °C)
                                   < 5 (at 50 °C)   (Wacker, 1994)

5.5   Density, g/cm³   1.31 (at 25 °C)   (Wacker, 1991)

5.6   Solubility in water   Reacts with water
                                                      (Wacker, 1991)

5.7   Solubility in organicsolvents   Soluble in aliphatic and aromatic
                                        chlorinated hydrocarbons
                                                            (Wacker, 1991)
                                        Soluble in chloroform, benzene,
                                        ether, perchloroethylene
                                                            (HSDB, 1992)

5.8   Solubility in fat   No information available

| 5.9 | pH-value | No information available |
| 5.10 | Conversion factor | $1\ ml/m^3$ (ppm) $\triangleq$ $8.63\ mg/m^3$ |

5.9   pH-value

No information available

5.10 Conversion factor

$1\ ml/m^3$ (ppm) $\triangleq 8.63\ mg/m^3$
$1\ mg/m^3 \triangleq 0.12\ ml/m^3$ (ppm)
(at 1013 hPa and 25 °C)

6.   Uses

Raw material in the manufacture of silicon resins (Wacker, 1991).

7.   Experimental results

7.1   Toxicokinetics and metabolism

No information available.

7.2   Acute and subacute toxicity

An oral LD_{50} value of 2.39 (2.19–2.61) g/kg body weight was determined in male Carworth-Wistar rats (body weight 90–120 g, 5 per group), following gastric intubation and a 14-day observation period (Smyth et al., 1959).

The dermal LD_{50} value for the rabbit (groups of 4 male New Zealand giant albinos, body weight 2.5–3.5 kg) was 0.89 (0.61–1.31) ml/kg body weight ($\triangleq$ 1166 (800–1716) mg/kg body weight). The test substance was applied occlusively to the closely clipped skin (1/10th of the body surface) for 24 hours, with a follow-up period of 14 days (Smyth et al., 1959). In a range-finding inhalation study, 6 male albino rats survived exposure to the concentrated vapour for 8 hours (no further details; Smyth et al., 1959).

An LC_{50} value of 0.33 mg/l was reported in mice (strain and sex not specified, 10 per group, body weight 20–23 g) following a 2-hour exposure to trichlorophenylsilane in a sealed chamber. Breathlessness was evident at lower concentrations (0.12–0.3 mg/l), and mucosal irritation and agitated motor activity, lying on the stomach, spreading of hind limbs, convulsions, asphyxia, cyanosis were also reported. The mice that died during the exposure period or in the subsequent 14-day observation period showed an excess of fluid in the lungs and brain, an effect which was attributed to increased permeability of the blood vessel walls. Other effects included swelling and proliferation of the reticulo-endothelial elements in the liver and, in those mice that died 3–14 days after exposure, parenchymatous degeneration in the liver, swelling of the bronchial epithelia, and in some cases signs of catarrhal pneumonia. The kidneys, spleen and heart appeared normal on microscopic examination, but changes were observed in the brain (Korlyakova, 1961).

An intravenous LD_{50} value of 100 mg/kg body weight was reported in mice (no further details known; US Army).

7.3　Skin and mucous membrane effects

Trichlorophenylsilane induced skin necrosis within 24 hours of application of the neat liquid (0.01 ml) to the clipped skin of 5 albino rabbits (strain and sex not specified; Smyth et al., 1959). A similar effect was reported (no experimental details known) by Union Carbide (1986).

Severe skin reactions were also reported by Korlyakova (1961). 3 drops of the chemical applied to the shaven skin of rabbits (number, strain and sex not specified) for 3 minutes resulted in the formation of a white patch at the site of contact, an inflammatory reaction, and subsequent development of a sore which took $1–1^{1}/_{2}$ months to heal; 2 drops applied to the interior surface of the ear for 5 minutes produced immediate hyperaemia of the skin, which changed into ischaemia after 5 minutes, a sharp inflammatory reaction at the application site on the following day, hyperaemia and oedema of the ear, and subsequent ulceration of the skin; the site had recovered after 2 weeks.

Severe skin irritation has been reported in rabbits given a 24-hour application of 5 mg (no further details known; Marhold, 1986).

The instillation of 0.005 ml neat trichlorophenylsilane into the rabbit eye caused severe burns, graded 9 on a 10-point scale of increasing irritation (no further details; Smyth et al., 1959). The same grading was reported (no experimental details known) by Union Carbide (1986).

Similarly, according to a study by Korlyakova (1961), a single drop of the chemical placed on the cornea of rabbits (number, strain and sex not specified) caused immediate desquamation of the epithelium and clouding of the cornea. On the third day, necrosis of the eyelid mucosa and diffuse clouding of the cornea were seen, and by day 14 there was scarring of the eyelid mucosa, ulceration and distortion of the eyelids, loss of the eyelashes, persistent corneal clouding and binding of the cornea and the conjunctiva of the lids. When the experiment was repeated but with immediate rinsing of the eye after treatment, corneal clouding and inflammation were again observed.

In studies involving exposure to trichlorophenylsilane vapour, signs of mild eye irritation were reported in mice exposed for 2 hours at vapour concentrations below 0.05 mg/l, and clear irritant effects were evident at higher concentrations (0.12–0.8 mg/l) with inflammation of the mucous membranes of the eyes on the third or fourth day after exposure (Korlyakova, 1961).

Severe eye irritation was reported in rabbits exposed for 3 minutes at a vapour concentration of 6 mg/l (no further details known; Union Carbide, 1986).

7.4   Sensitization

No information available.

7.5   Subchronic and chronic toxicity

No information available.

7.6   Genotoxicity

7.6.1 In vitro

No information available.

7.6.2 In vivo

No information available.

7.7   Carcinogenicity

No information available.

7.8   Reproductive toxicity

No information available.

7.9   Effects on the immune system

No information available.

7.10 Neurotoxicity

No information available.

7.11 Other effects

No information available.

8.   Experience in humans

No specific health effects have been observed in the 150 workers of a organochlorosilane- and organosilane production plant in the last 20–25 years from the occupational medical department of the firm (Hüls, 1993).

No cases of accidental exposure to trichlorophenylsilane have been identified. However, the chlorosilanes as a group are considered to be hazardous on acute exposure, as hydrochloric acid is formed on their hydrolysis (Sax, 1975). Different data banks report on health risks after acute exposure of trichlorophenylsilane. The vapour is likely to be severely irritating to the eyes, nose and throat; skin or eye contact with the liquid will result in severe burns, and ingestion is likely to cause severe burns of the mouth, oesophagus and stomach (CHRIS, 1991; HSDB, 1992). The basis, however, for these statements is not known; possibly they are derived from the results of animal experiments.

9.   Threshold limit values

No information available.

References

CHRIS (Chemical Hazard Response Information System)
The United States Coast Guard, US Department of Transportation
SilverPlatter, Chem-Bank, März 1991

HSDB, Hazardous Substances Data Bank
US National Library of Medicine
SilverPlatter, Chem-Bank, February 1992

Hüls AG, Werksärztlicher Dienst, Werk Rheinfelden
Written communication to BG Chemie, 10.03.1993

Korlyakova, E.A.
Toxicology of certain monomers of organosilicon compounds (trichloromethylsilane, dichlorodimethylsilane, trichloroethylsilane,
trichlorophenylsilane)
Toksikol. Nov. Prom. Khim. Veshchestv., 3, 23–33 (1961)

Marhold, J.V.
Prehled Prumyslove Toxikol. Org. Latky, p. 1228–1229 (1986)
Cited in: RTECS (1992)

RTECS (Registry of Toxic Effects of Chemical Substances)
Silane, trichlorophenyl-, RTECS-Nr. VV6650000
produced by NIOSH (National Institute for Occupational Health
and Safety) (1992)

Sax, N.I.
Dangerous Properties of Industrial Materials
4th edn., p. 554
Van Nostrand Reinhold Company, New York (1975)

Smyth, H.F., jr., Carpenter, C.P., Weil, C.S., Pozzani, U.C.
Range-finding toxicity data, list V
A.M.A. Arch. Ind. Hyg. Occup. Med., 10, 61–68 (1959)

Union Carbide Corporation
Rabbit eye and skin injury testing on seven silica compounds
NTIS/OTS0000469-0, FYI-OTS-0386–0469 (1986)

US Army Armament Research and Development Command, Chemical Systems Laboratory, NIOSH Exchange Chemicals. NX# 04051.
(Aberdeen Proving Ground, MD 21010) (Undated)
Cited in: RTECS (1992)

Wacker-Chemie GmbH
Grunddatensatz für Großstoffe - Trichlorphenylsilan (1991)

Wacker-Chemie GmbH
Written communication to BG Chemie, 13.06.1994

4-Nitro-1,2-dimethylbenzene

1. Summary and assessment

4-Nitro-1,2-dimethylbenzene is of very low acute toxicity on oral administration (LD_{50} rat oral, between 2500 and 4000 mg/kg body weight in males, 2600 mg/kg body weight in females).

On repeated administration in the diet of 0, 200, 1000 or 5000 ppm for 5 days or of 0, 60, 300 or 1500 ppm for 28 days (equivalent to 20, 100 and 500 mg/kg body weight/day and 6, 30 and 150 mg/kg body weight/day, respectively), reduced weight gain and decreased feed intake were seen at the top dose (5000 ppm) in the 5-day study. The weights of the liver, kidneys and spleen were reduced, but no corresponding histological changes were found. No treatment-related effects were seen in the 28-day study at doses of up to 1500 ppm. The no effect level was therefore given as 1500 ppm (equivalent to ca. 150 mg/kg body weight/day).

The substance causes transient, very slight inflammation of the rabbit skin and eyes and is evaluated as not irritating to either the skin or the eyes, in accordance with EC classification criteria.

4-Nitro-1,2-dimethylbenzene is not mutagenic in vitro in either *Salmonella typhimurium* or *Escherichia coli* or in the chromosome aberration test in Chinese hamster V79 cells.

2. Name of substance

2.1	Usual name	4-Nitro-1,2-dimethylbenzene
2.2	IUPAC-name	4-Nitro-1,2-dimethylbenzene
2.3	CAS-No.	99–51–4
2.4	EINECS-No.	202–761–6

3.   Synonyms, common and trade names

1,2-Dimethyl-4-nitrobenzene
4-Nitro-1,2-dimethylbenzol
4-Nitro-o-xylene
Nitro-o-xylene (412)
Nitro-o-xylene, asym.

4.   Structural and molecular formulae

4.1   Structural formula

4.2   Molecular formula $C_8H_9NO_2$

5.   Physical and chemical properties

5.1   Molecular mass, g/mol   151.16

5.2   Melting point, °C   28–30 (Thiem et al., 1979)

5.3   Boiling point, °C   258 (at 1013 hPa)
(Thiem et al., 1979)

5.4   Vapour pressure, hPa   No information available

5.5   Density, g/cm³   1.139 (at 20 °C)
(Thiem et al., 1979)

5.6   Solubility in water   Insoluble (Thiem et al., 1979)

5.7   Solubility in organic solvents   Dissolves readily in most organic solvents (Thiem et al., 1979)

5.8   Solubility in fat   No information available

5.9   pH-value   –

5.10 Conversion factor	$1\ ml/m^3$ (ppm) $\triangleq 6.17\ mg/m^3$
	$1\ mg/m^3 \triangleq 0.16\ ml/m^3$ (ppm)
	(at 1013 hPa and 25 °C)

6. Uses

In the manufacture of 3,4-xylidine, from which riboflavin (vitamin B_2) is manufactured in a multi-stage process (Thiem et al., 1979).

7. Experimental results

7.1 Toxicokinetics and metabolism

No information available.

7.2 Acute and subacute toxicity

The acute oral toxicity of 4-nitro-1,2-dimethylbenzene was studied in Wistar rats (5 males and 5 females/dose). The LD_{50} value in males was between 2500 and 4000 mg/kg body weight, while that in females was 2600 mg/kg body weight. Impairment of breathing, coordination and reflexes was seen in the rats. The signs of toxicity were reversible from day 7 of the observation period. Post-mortems revealed discoloration of the liver and spleen and reddening of the mucous membrane of the gastrointestinal tract in the animals that died during the study. The lungs were hyperaemic. No macroscopic effects were evident in the rats killed after the 14-day observation period (Hoechst, 1987 a).

In a 5-day range-finding study carried out prior to a 28-day feeding study, groups of 5 male and 5 female Wistar rats, aged 3 to 4 weeks were given 0, 200, 1000 or 5000 ppm 4-nitro-1,2-dimethylbenzene in the diet (purity 99 %; equivalent to 0, 20, 100 and 500 mg/kg body weight/day, respectively). The homogeneous distribution and stability of the test substance in the feed complied with requirements. All of the rats survived. There were no differences in general behaviour between the treated and control rats. At the top and middle dose levels, body weights were lower than those of the controls in both sexes, although this effect was only statistically significant at the top dose. Feed intake was also reduced in both sexes in these dose groups. The erythrocyte count, haemoglobin concentration and haematocrit value were increased in males in the medium and high dose groups. These effects were not seen in female rats. The absolute liver and kidney weights in the males and the absolute and relative spleen weights in the females were reduced at the top dose. Post-mortems revealed no treatment-related effects compared with the controls (TNO, 1991).

In the subsequent 28-day feeding study, groups of 5 male and 5 female rats received 0, 60, 300 and 1500 ppm 4-nitro-1,2-dimethylbenzene in the diet (purity 99%; equivalent to 0, 6, 30 and 150 mg/kg body weight/day). A further 5 male and 5 female rats were included at 0 and 1500 ppm and kept for observation after the study. The concentration of the test substance in the diet and its distribution complied with requirements. No signs of toxicity were seen during the study and all of the rats survived. There were no differences in body weight gain or feed and water intake between the treated and control rats. No differences from the controls were seen in red and white blood cell counts or blood clotting. In the clinical-chemistry tests, only the plasma creatinine level was significantly increased in the males at 1500 ppm at the end of the study, but it was still within the range of historical control values and the increase was reversible within the 14-day observation period. Urinary parameters did not differ from those of the controls. There were no changes in the absolute and relative organ weights of the liver, kidneys, adrenal glands, testes or spleen compared with the controls. No effects that could be attributed to administration of the substance were evident on histological examination of the adrenal glands, heart, kidneys, liver or spleen. The no effect level derived from this study was 1500 ppm (equivalent to ca. 150 mg/kg body weight/day; TNO, 1991).

7.3   Skin and mucous membrane effects

The acute skin irritancy of 4-nitro-1,2-dimethylbenzene (> 99% pure) was studied after application of the substance to the clipped dorsal skin of New Zealand albino rabbits in accordance with OECD guideline no. 404 (4-hour semi-occlusive exposure to the neat substance). One hour after removal of the patch, barely perceptible to clear localized erythema and barely perceptible oedema occurred. In 1 of the 3 rabbits, a large area of yellow discoloration was seen. No erythema was perceptible 24 hours after removal of the patch. All of the irritant effects had cleared up 48 hours after removal of the patch. In accordance with the classification criteria in the guidelines contained in 83/467/EEC, the product was evaluated as not irritating (Hoechst, 1987 b).

The acute eye irritancy of > 99% pure 4-nitro-1,2-dimethylbenzene was studied in New Zealand albino rabbits in accordance with OECD guideline no. 405 (100 mg/eye, neat). One hour after instillation, the conjunctiva showed a diffuse crimson to severe red colouration and slight swelling that was greater than normal. The iris was reddened in 2 of the 3 rabbits. Some of the conjunctival blood vessels showed marked hyperaemia in 2 of the rabbits 24 hours after instillation. All of the irritant effects were reversible from 48 hours. The product was evaluated as not irritating in accordance with the classification criteria in the guidelines in 83/467/EEC (Hoechst, 1987 c).

7.4   Sensitization

No information available.

7.5   Subchronic and chronic toxicity

No information available.

7.6   Genotoxicity

7.6.1 In vitro

The mutagenic activity of 4-nitro-1,2-dimethylbenzene (100% pure) was studied in the plate incorporation test with and without S9-mix (from Aroclor 1254-induced rat liver) at concentrations of from 3.3 to 1000 µg/plate in the *Salmonella typhimurium* strains TA 1535, TA 1537, TA 1538, TA 98 and TA 100 and in *Escherichia coli* WP2. Slight and severe bacteriotoxicity were evident (in the pre-experiment) at 1000 and 5000 µg/plate, respectively. The revertant count was in the region of that of the controls. Thus 4-nitro-1,2-dimethylbenzene was not mutagenic in this study (CCR, 1988 a).

In a further study in the Salmonella/microsome test in TA 100 and TA 98, there were indications of a dose-dependent mutagenic effect with the addition of S9-mix (from Kanechlor 500-induced rat liver), although the increase in the revertant count was not greater than 2-fold. No increase in revertants was seen in strains TA 98 and TA 100 without S9-mix (no further details; Kawai et al., 1987). In accordance with internationally recognized evaluation criteria, 4-nitro-1,2-dimethylbenzene was not mutagenic in this study.

In a review paper, 4-nitro-1,2-dimethylbenzene was reported to be mutagenic in the *Salmonella typhimurium* strain TA 100 with S9-mix (from Kanechlor KC-400-induced rat liver). There were no indications of mutagenic activity in this strain without S9-mix or in strain TA 98 with or without S9-mix (no further details; Nohmi et al., 1984).

4-Nitro-1,2-dimethylbenzene (100% pure) was studied in a chromosome aberration test in Chinese hamster V79 cells in vitro with and without S9-mix (from Aroclor 1254-induced rat liver). Slides were prepared 7, 18 and 28 hours after the beginning of the study. The following concentrations were employed:

	with S9-mix (µg/ml)	without S9-mix (µg/ml)
7 hours	200	200
18 hours	10, 200, 250	10, 200, 300
28 hours	300	300

The test substance did not induce structural chromosome aberrations either with or without a metabolizing system. 4-Nitro-1,2-dimethylbenzene was therefore evaluated as not mutagenic in this study (CCR, 1988 b).

7.6.2 In vivo

No information available.

7.7   Carcinogenicity

No information available.

7.8   Reproductive toxicity

No information available.

7.9   Effects on the immune system

No information available.

7.10   Neurotoxicity

No information available.

7.11   Other effects

No information available.

8.   Experience in humans

No information available.

9.   Threshold limit values

No information available.

References

CCR (Cytotest Cell Research GmbH & Co KG)
Salmonella typhimurium and *Escherichia coli* reverse mutation assay with 4-Nitro-1,2-dimethylbenzol
Unpublished report, CCR project 116111 (1988 a)
On behalf of BG Chemie

CCR (Cytotest Cell Research GmbH & Co KG)
Chromosome aberration assay in Chinese hamster V79 cells in vitro
with 4-Nitro-1,2-dimethylbenzol
Unpublished report, CCR project 116122 (1988 b)
On behalf of BG Chemie

Hoechst AG, Pharma Forschung Toxikologie und Pathologie
Nitro-o-xylol asym. - Prüfung der akuten oralen Toxizität an der
männlichen und weiblichen Wistar-Ratte
Unpublished report no. 87.0212 (1987 a)

Hoechst AG, Pharma Forschung Toxikologie und Pathologie
Nitro-o-xylol asym. - Prüfung auf Hautreizung am Kaninchen
Unpublished report no. 87.0109 (1987 b)

Hoechst AG, Pharma Forschung Toxikologie und Pathologie
Nitro-o-xylol asym. - Prüfung auf Augenreizung am Kaninchen
Unpublished report no. 87.0134 (1987 c)

Kawai, A., Goto, S., Matsumoto, Y., Matsushita, H.
Mutagenicity of aliphatic and aromatic nitro compounds
Jpn. J. Ind. Health (Sangyo Igaku), 29, 34–54 (1987)

Nohmi, T., Yoshikawa, K., Nakadate, M., Miyata, R., Ishidate, M., jr.
Mutations in *Salmonella typhimurium* and inactivation of *Bacillus
subtilis* transforming DNA induced by phenylhydroxylamine derivatives
Mutat. Res., 136, 159–168 (1984)

Thiem, K.W., Sewekow, B., Kiel, W., Handschuh, V., Freese, H.,
Schimpf, R., Vagt, H., Bunge, W.
Nitroverbindungen, aromatische
In: Ullmanns Encyklopädie der technischen Chemie
4th ed., vol. 17, p. 383–416
Verlag Chemie, Weinheim (1979)

TNO Nutrition and Food Research, Zeist, Netherlands
Short-term oral toxicity studies with 4-Nitro-1,2-dimethylbenzol in

rats: - range-finding (5-day) study, - sub-acute (28-day) study with (14-day) recovery study
Unpublished report no. V88.464 (1991)
On behalf of BG Chemie

Comparison of the toxic effects of five different nitro-dimethylbenzene-isomers (nitroxylenes)

4-Nitro-1,3-dimethylbenzene (CAS-No. 89–87–2)
2-Nitro-1,3-dimethylbenzene (CAS-No. 81–20–9)
4-Nitro-1,2-dimethylbenzene (CAS-No. 99–51–4)
3-Nitro-1,2-dimethylbenzene (CAS-No. 83–41–0)
2-Nitro-1,4-dimethylbenzene (CAS-No. 89–58–7)

The nitroxylenes that have been studied are of low acute oral toxicity. 2-Nitro-1,4-dimethylbenzene has not been studied. The LD_{50}-values are around or above 2000 mg/kg body weight except the case of 4-nitro-1,3-dimethylbenzene in female rats (1690 mg/kg body weight). Similar signs of toxicity are observed for all of the substances (reduced movement, respiratory depression, inhibition of reflexes, stupefaction, watering of the eyes). Post-mortems on animals killed during studies or after the observation period have revealed no treatment-related effects on the organs.

None of the nitroxylenes studied are irritating to the skin or eyes of rabbits as defined by the guidelines contained in 83/467/EEC. 2-Nitro-1,4-dimethylbenzene has not been studied.

The administration of 4-nitro-1,3-dimethylbenzene, 2-nitro-1,3-dimethylbenzene, 4-nitro-1,2-dimethylbenzene or 3-nitro-1,2-dimethylbenzene to rats for 5 days at dietary concentrations of up to 5000 ppm produces very similar results for all of the substances at this dose. A reduction in body weight, reduced feed intake and an increase in the relative liver weight occur only at the high dose level. No pathological effects have been seen at post-mortem.

In 28-day feeding studies conducted in accordance with OECD guideline no. 407, 4-nitro-1,3-dimethylbenzene and 2-nitro-1,3-dimethylbenzene were added to the diet at concentrations of 0, 100, 600 and 3000 ppm (equivalent to 0, 10, 60 and 300 mg/kg body weight/day, respectively), 4-nitro-1,2-dimethylbenzene at concentrations of 0, 60, 300 and 1500 ppm (equivalent to 0, 6, 30 and 150 mg/kg body weight/day, respectively) and 3-nitro-1,2-dimethylbenzene at concentrations of 0, 100, 500 and 2500 ppm (equivalent to 0, 10, 50 and 250 mg/kg body weight/day). Effects on the blood count (decreases in haemoglobin and erythrocytes) and on liver function (effects on the liver enzymes and a reduction in total protein and albumin) were observed with all 4 substances at their respective highest doses, particularly with 4-nitro-1,3-dimethylbenzene. These effects were reversible by the end of the observation period (14 days), as were the increases in absolute and relative weights of the liver, spleen and adrenal glands seen with 4-nitro-1,3-dimethylbenzene and 2-nitro-1,3-dimethylbenzene at the end of the studies. Histopathological examination of the organs revealed no treatment-related effects. There were no unequivocal indications of a specific target organ. The no effect levels were 600 ppm (ca. 60 mg/kg body weight/day) for 4-nitro-1,3-dimethylbenzene and 2-nitro-1,3-dimethylbenzene and 500 ppm (ca. 50 mg/kg body weight/day) for 3-nitro-1,2-dimethylbenzene. The no effect level for 4-nitro-1,2-dimethylbenzene was given as 1500 ppm (ca. 150 mg/kg body weight/day).

The results of genotoxicity studies on the five nitroxylenes are summarized in the table below. In the Salmonella/microsome test, 4-nitro-1,3-dimethylbenzene is mutagenic in strain TA 100 with and without metabolic activation and strain TA 1535 without metabolic activation while 2-nitro-1,4-dimethylbenzene is mutagenic in strain TA 100 with metabolic activation. In mammalian cells (HPRT test in Chinese hamster V79 cells with and without metabolic activation), 4-nitro-1,3-dimethylbenzene (the only nitroxylene that has been studied in this test) is not mutagenic. 4-Nitro-1,3-dimethylbenzene, 2-nitro-1,3-dimethylbenzene and 3-nitro-1,2-dimethylbenzene are clastogenic in the presence of S9-mix in the chromosome aberration test in Chinese hamster V79 cells in vitro, whereas 4-nitro-1,2-dimethylbenzene is not. Without metabolic activation, the 4 compounds do not cause chromosome aberrations in this in vitro test. In vivo, 4-nitro-1,3-dimethylbenzene is not clastogenic in the Chinese hamster (no other nitroxylenes have been studied in this test).

Thus, based on the available results, the nitroxylenes show no fundamental differences in their toxicity profiles. With the exception of 4-nitro-1,2-dimethylbenzene, there is evidence of a mutagenic potential in all the other tested nitroxylenes, particularly 4-nitro-1,3-dimethylbenzene. No structure-activity relationship between the nitroxylenes can be discerned with regard to mutagenic activity.

Table. Results of genotoxicity studies on five nitro-dimethylbenzene-isomers

Substance		Salmonella/microsome test (plate incorporation test)						HPRT V79 cells Chinese hamster	C in vitro V79 cells Chinese hamster	C in vivo Chinese hamster
		TA 98	TA 100	TA 1535	TA 1537	TA 1538	E. coli			
4-Nitro-1,3-dimethylbenzene	+ S9	–	+	–	–	–	–	–	+	–
	– S9	–	+	+	–	–	–	–	–	
2-Nitro-1,3-dimethylbenzene	+ S9	–	–	–	–	–	–	n.s.	+	n.s.
	– S9	–	–	–	–	–	–	n.s.	–	
4-Nitro-1,2-dimethylbenzene	+ S9	–	–	–	–	–	–	n.s.	–	n.s.
	– S9	–	–	–	–	–	–	n.s.	–	
3-Nitro-1,2-dimethylbenzene	+ S9	–	–	–	–	–	–	n.s.	+	n.s.
	– S9	–	–	–	–	–	–	n.s.	–	
2-Nitro-1,4-dimethylbenzene	+ S9	–	+	n.s.	n.s.	n.s.	n.s.	n.s.	n.s.	n.s.
	– S9	–	–	n.s.	n.s.	n.s.	n.s.	n.s.	n.s.	

C	Chromosome aberration test
+ S9	with metabolic activation system (S9–mix)
– S9	without metabolic activation system (S9–mix)
+	mutagenic
–	not mutagenic
n.s.	not studied

Isobornyl acetate

1. Summary and assessment

Isobornyl acetate is rapidly absorbed through the skin, reaching its maximum concentration in the blood after just 10 minutes.

The acute toxicity of isobornyl acetate is very low (LD_{50} rat oral >10 g/kg body weight, LD_{50} rabbit dermal >20 g/kg body weight).

Isobornyl acetate is irritating to the skin, while in the eye, transient, reversible reddening and swelling of the conjunctiva occur, effects which nevertheless may be evaluated as being not irritating in accordance with the guidelines contained in the directive 83/467/EEC.

Inhalation of isobornyl acetate for 4 or 8 weeks (concentration $7.9{\times}10^{-9}$ M, equivalent to 1.551 g/m^3; duration of exposure/day not specified) leads to effects on the olfactory epithelium, the significance of which is, however, unclear.

Repeated oral administration of 270 or 90 mg isobornyl acetate/kg body weight/day by gavage for 13 weeks has been shown to be nephrotoxic to rats. Liver damage was observed in this study at 270 mg/kg body weight/day. The no effect level in rats for daily administration by gavage for 13 weeks is 15 mg/kg body weight/day.

Isobornyl acetate is not mutagenic in the Salmonella/microsome test in the *Salmonella typhimurium* strains TA 1535, TA 1537, TA 1538, TA 98 and TA 100 at concentrations of up to 500 µg/plate, with or without metabolic activation. Isobornyl acetate does not induce chromosome damage in the micronucleus test in the mouse.

No maternal toxicity, embryotoxicity or teratogenic effects have been caused by the oral administration of 1000 mg isobornyl acetate/kg body weight/day to pregnant rats from day 7 to day 16 of pregnancy.

Isobornyl acetate induces microsomal enzymes of the liver both in vitro and in vivo.

Observations in humans (after exposure at the workplace and in 25 volunteers in a maximization test) have shown that the substance does not induce sensitization.

2.   Name of substance

2.1	Usual name	Isobornyl acetate
2.2	IUPAC-name	2-Acetoxy-1,7,7-trimethyl-bicyclo[2.2.1]heptane
2.3	CAS-No.	125–12–2
2.4	EINECS-No.	204–727–6

3.   Synonyms, common and trade names

Bicyclo-[2.2.1]-heptane-2-ol-1,7,7-trimethylacetate
Acetic acid isobornyl ester
Isoborneol acetate
Isobornylacetat extra
Isobornylacetat
Pichtosin
1,7,7-Trimethyl-bicyclo[2.2.1]-heptane-2-ol-exo-acetate

4.   Structural and molecular formulae

4.1   Structural formula

4.2   Molecular formula $C_{12}H_{20}O_2$

5.   Physical and chemical properties

5.1	Molecular mass, g/mol	196.29
5.2	Melting point, °C	No information available
5.3	Boiling point, °C	ca. 215 (Hoechst, 1989) 102–103 (at 16–17 hPa) (Bauer et al., 1988)
5.4	Vapour pressure, hPa	0.13 (at 20 °C) 13.3 (at 100 °C) (Hoechst, 1989)
5.5	Density, g/cm^3	0.98 (at 20 °C) (Hoechst, 1989)
5.6	Solubility in water	<1 g/l (Hoechst, 1989)
5.7	Solubility in organic solvents	Dissolves well in most organic solvents (Hoechst, 1988 a)
5.8	Solubility in fat	No information available
5.9	pH-value	–
5.10	Conversion factor	1 ml/m^3 (ppm) $\triangleq$ 8.01 mg/m^3 1 mg/m^3 $\triangleq$ 0.12 ml/m^3 (ppm) (at 1013 hPa and 25 °C)

6.   Uses

As a fragrance, e.g. in bath products, deodorants, soaps, air fresheners and other perfumery products (Hoechst, 1988 a); in the manufacture of camphor (Bauer et al., 1988).

7.   Experimental results

7.1   Toxicokinetics and metabolism

The cutaneous absorption of tritium-labelled isobornyl acetate was studied in male dd mice. The labelled substance was added to a liquid foam bath to give a concentration of 3.18 g/ml bath water. The foam bath also contained unlabelled camphor, menthol, α-pinene and limonene in addition to isobornyl acetate. The concentration of the substance corresponded to that obtained when the product was added to a full bath in accordance with the recommendations of the foam bath manufacturer. 3 mice had their backs shaved and depilated. The bath liquid was applied to a 3 cm^2 area by means of a cylinder applied to the skin. After exposure for 5, 10, 20, 30 and 40 minutes, 0.1 ml blood was taken from the tail vein and its radioactivity determined. After 5 minutes, 50% of the maximum value had already been reached; the maximum, ca. 60 ng isobornyl acetate/ml, was reached after 10 minutes. There was then a rapid fall in the concentration, which declined to about 15 ng/ml after 40 minutes. In a second study, the size of the application site exposed to the bath liquid was altered to 1.5, 3 or 6 cm^2. 3 mice were used per study. The bath liquid was applied for 10 minutes, after which the radioactivity in the blood was measured. The blood concentration increased linearly with the absorption surface and reached a concentration of 120 ng isobornyl acetate/ml when a 6 cm^2 application site was used (Schäfer and Schäfer, 1982).

7.2   Acute and subacute toxicity

The oral LD_{50} was reported as >10 g/kg in rats, the dermal LD_{50} as >20 g/kg in rabbits (no further details; Fogleman, 1970).

In order to study whether fragrances led to specific cell degeneration in the olfactory bulb, groups of 3 young Wistar rats (2 weeks old, 28 to 29 g, sex not specified) inhaled isobornyl acetate at a concentration of 7.9×10^{-9} M (1.551 mg/m^3) for 4 weeks (final weight 107 g) or 8 weeks (final weight 220 g; no data on duration of exposure/day). The rats were placed in cylindrical cages, through which filtered air was passed; the air was mixed with isobornyl acetate. At the end of the exposure period, the rats were perfused with buffered saline and fixed in a mixture of formaldehyde and glutaraldehyde, and the olfactory bulb was then stained and examined microscopically. Compared with the control rats (same litter, filtered air), isobornyl acetate had clear effects on the granulated cells in the central area of the olfactory bulb (no details of clinical findings; Pinching and Døving, 1974).

7.3   Skin and mucous membrane effects

The skin irritancy of isobornyl acetate (purity 93.5 to 95.5%, 0.5 ml) was studied after application of the substance to a 2.5×2.5 cm area of the clipped dorsal skin of 3 New Zealand white rabbits. The surface was covered for 4 hours with adhesive plaster and a semi-occlusive bandage, after which the test substance was washed off. After 60 minutes, mild reddening and mild oedema were visible, 24 to 72 hours after the application, clear localized erythema was seen, and after 7 days, the skin was dry and flaky. The substance was evaluated as irritating to the skin (Hoechst, 1988 a).

Neat isobornyl acetate was only slightly irritating to the intact or abraded skin of rabbits after occlusive exposure for 24 hours (no further details; Fogleman, 1970).

The eye irritancy of isobornyl acetate was studied in 3 rabbits (New Zealand white). The test substance (0.1 ml, purity 93.5 to 95.5%) was instilled into the conjunctival sac. The eyes were examined 1, 24, 48 and 72 hours after instillation. After 1 hour, marked redness and swelling of the conjunctiva was visible, after 24 hours slight redness and swelling could still be observed, while after 72 hours the conjunctiva appeared normal again. There were no effects on the cornea (examination with fluorescein). The substance was evaluated as not irritating in accordance with the guidelines contained in the directive 83/467/EEC (Hoechst, 1988 b).

7.4   Sensitization

No information available (see also Sect. 8).

7.5   Subchronic and chronic toxicity

CFE rats (15 males and 15 females/group, SPF-maintained) were given isobornyl acetate (purity 97%) in corn oil by gavage daily for 13 weeks at doses of 0, 15, 90 or 270 mg/kg body weight/day. Additional groups of 5 males and 5 females/dose were treated with the substance for up to 6 weeks. The following parameters were determined: body weight gain, feed and water consumption (weekly), urinalysis during the final week of the study (week 6 or week 13), and concentrating ability of the kidneys. The rats were killed 24 hours after administration of the last dose and blood was taken for analysis of the following parameters: haemoglobin, haematocrit, numbers of erythrocytes, reticulocytes and leukocytes, differential blood count, urea, glucose, total protein, albumin, GOT (aspartate aminotransferase), GPT (alanine aminotransferase) and LDH (serum lactate dehydrogenase). The rats were dissected and the following organs weighed: brain, thyroid gland, pituitary gland, heart, liver, spleen, adrenal glands, kidneys, testes; with the stomach, small intestine and caecum also being weighed after 13 weeks. These organs, as well as the lungs, lymph nodes, thymus, bladder, colon, rectum, pancreas, uterus and muscles were examined histologically. None of the rats died during the study. There were no differences in behaviour, body weight gain, feed consumption or the results of the clinical chemistry and haematological tests between the treated rats and the controls. The concentrating ability of the kidneys was reduced and drinking water consumption increased in the males given 270 mg/kg/day. No effects on renal concentrating ability were detected at the lower doses. At 270 mg/kg, a significant increase in cells in the urine was observed from week 6, an effect which also occurred in the 90 mg/kg group in week 13. The relative kidney weight was increased at 270 mg/kg from week 6 in males and in both sexes after week 13. Histologically, the kidneys of rats at the highest dose level (males and females) showed tubular degeneration, and vacuolization of the tubule cells was seen in the males. The relative liver weight was increased in both sexes at 270 mg/kg after 13 weeks, and the males also showed vacuolization of the epithelium of the intrahepatic bile duct. The absolute and relative caecum weights were increased in both sexes at this dose level. The dose of 15 mg/kg body weight/day had no effects on the parameters studied when administered for 13 weeks (Gaunt et al., 1971).

7.6   Genotoxicity

7.6.1 In vitro

Isobornyl acetate (dissolved in DMSO; no information on purity of the substance provided) was studied in the Salmonella/microsome test in strains TA 1535, TA 1537, TA 1538, TA 98 and TA 100 at concentrations of from 10 to 500 µg/plate with and without S9-mix (from Aroclor 1254-induced rat livers) in accordance with OECD guideline no. 471. No mutagenic activity was detected (Hoechst, 1988 c).

7.6.2 In vivo

In a micronucleus test which was carried out in accordance with OECD guideline no. 474, groups of 5 female and 5 male NMRI mice were given a single dose of 2000 mg isobornyl acetate/kg body weight (the highest non-lethal dose) by gavage, and compared with a positive control group (50 mg/kg Endoxan) and an untreated control group. The bone marrow was removed from both femurs of the mice 24, 48 or 72 hours after administration of the substance. The number of micronucleated cells in 1000 polychromatic and 1000 normochromatic erythrocytes/animal was determined, as was the ratio of polychromatic to normochromatic erythrocytes. Neither parameter was affected by administration of the substance, and isobornyl acetate thus had no adverse effects on the chromosomes in the micronucleus test (Hoechst, 1991).

7.7   Carcinogenicity

No information available.

7.8   Reproductive toxicity

In a limit test carried out in accordance with OECD guideline no. 414, 20 female Wistar rats (65 to 70 days old, weight ca. 191 g) were given 1000 mg isobornyl acetate/kg body weight/day in sesame oil by gavage from day 7 to day 16 of pregnancy. Doses of 270, 500 and 1000 mg/kg body weight/day were shown not to cause maternal or embryo toxicity in a preliminary study. On day 21 of pregnancy, the uteri were opened up and examined macroscopically, and the numbers of live and dead foetuses, resorption sites and placentae, as well as the number of corpora lutea in the ovaries, were determined. The diameter of the resorbed foetuses and the placental weights were also measured. The foetuses and the dams were subjected to a thorough macroscopic examination and the

viscera and skeletons of the foetuses were evaluated microscopically. The substance did not have pathological effects on either the
dams or the foetuses, and thus showed no embryotoxic or teratogenic activity (Hoechst, 1992).

7.9   Effects on the immune system

No information available.

7.10 Neurotoxicity

No information available.

7.11 Other effects

Isobornyl acetate induced microsomal enzymes in the liver. 4 adult
male Sprague-Dawley rats were given 500 or 1000 mg isobornyl
acetate/kg body weight/day (purity unspecified) by intraperitoneal
injection for 3 days. Enzymes that were dose-dependently induced
in the liver microsomes included cytochrome P-450 (up to 2.5-fold
increase over the control value), cytochrome b_5 (1.8 fold), aminopyrine N-demethylase and ethylmorphine demethylase (up to 2.6- and
2.5-fold increase over the control values, respectively) and NADP
cytochrome C reductase (up to 2 times the control value). In vitro,
25 or 100 M isobornyl acetate dose-dependently increased the N-
methylation of aminopyrine in a rat liver microsome suspension
compared with the controls (Cinti et al., 1976).

8.   Experience in humans

Isobornyl acetate was tested on 25 volunteers in a maximization
test (modification of the method of Kligman, 1966). The substance
was applied at a concentration of 10% in vaseline. No sensitization
reactions occurred (Kligman, 1970).

No cases of sensitization to isobornyl acetate were observed in an
isobornyl acetate processing plant where dermal and inhalation
exposure could have occurred (Dragoco, 1991).

9.   Threshold limit values

No information available.

References

Bauer, K., Garbe, D., Surburg, H.
Flavors and fragrances
In: Ullmann's encyclopedia of industrial chemistry
5th edn., vol. A 11, p. 177
VCH Verlagsgesellschaft, Weinheim (1988)

Cinti, D.L., Lemelin, M.A., Christian, J.
Induction of liver microsomal mixed-function oxidases by volatile hydrocarbons
Biochem. Pharmacol., 25, 100–103 (1976)

Dragoco, Holzminden
Written communication to BG Chemie of 17.12.1991

Fogleman, R.W.
Report to RIFM (Research Institute for Fragrance Materials), 25 August 1970
Cited in: Fragrance raw materials monographs
Food Cosmet. Toxicol., 13, 552 (1975)

Gaunt, I.F., Agrelo, C.E., Colley, J., Lansdown, A.B.G., Grasso, P.
Short-term toxicity of isobornyl acetate in rats
Food Cosmet. Toxicol., 9, 355–366 (1971)

Hoechst AG, Pharma Forschung Toxikologie und Pathologie
Isobornylacetat-Extra, Prüfung auf Hautreizung am Kaninchen
Unpublished report no. 88.0190 (1988 a)

Hoechst AG, Pharma Forschung Toxikologie und Pathologie
Isobornylacetat-Extra, Prüfung auf Augenreizung am Kaninchen
Unpublished report no. 88.0248 (1988 b)

Hoechst AG, Pharma Forschung Toxikologie und Pathologie
EXT 8806 Isobornylacetate, reverse mutation assay in vitro, Ames
test
Unpublished report no. 88.0896 (1988 c)

Hoechst AG
DIN-Safety data sheet Isobornylacetat (1989)

Hoechst AG
Isobornylacetat, CAS-Nr. 125–12–2
Written communication to BG Chemie of 04.10.1990

Hoechst AG, Abteilung Toxikologie
Isobornylacetat-Extra, micronucleus test in male and female NMRI
mice after oral administration
Unpublished report no. 91.0220 (1991)

Hoechst AG, Pharma Entwicklung Toxikologie
Isobornylacetat extra, Prüfung auf embryotoxische Wirkung an
Wistar-Ratten bei oraler Verabreichung (Limit-Test)
Unpublished report no. 92.0243 (1992)

Kligman, A.M.
The identification of contact allergens by human assay. III. The maximization test: a procedure for screening and rating contact sensitiz-
ers
J. Invest. Dermatol., 47 (5), 393–409 (1966)

Kligman, A.M.
Report to RIFM (Research Institute for Fragrance Materials), 7 Octo-
ber 1970
Cited in: Fragrance raw materials monographs
Food Cosmet. Toxicol., 13, 552 (1975)

Pinching, A.J., Dving, K.B.
Selective degeneration in the rat olfactory bulb following exposure
to different odours
Brain Res., 82, 195–204 (1974)

Schäfer, R., Schäfer, W.
Die perkutane Resorption verschiedener Terpene - Menthol, Campher, Limonen, Isobornylacetat, α-Pinen - aus Badezusätzen
Arzneimittelforschung, 32 (1), 56–58 (1982)

Propargyl chloride

1. Summary and assessment

Propargyl chloride is acute toxic on oral or dermal exposure (LD_{50} rat oral ca. 165 mg/kg body weight).

In screening inhalation studies, propargyl chloride has been shown to be relatively toxic.

The sensory irritant effect of propargyl chloride is relatively low, with an RD_{50} value of from 1 to 2 mg/l.

Depending on the concentration, propargyl chloride is irritant to corrosive on contact with the skin or eye of the rabbit.

Propargyl chloride induces genetic mutations in the Salmonella/microsome test. Observations in the micronucleus test in the mouse have shown it to be neither a clastogen nor a spindle poison.

2. Name of substance

2.1	Usual name	Propargyl chloride
2.2	IUPAC-name	1-Chloro-2-propyne
2.3	CAS-No.	624–65–7
2.4	EINECS-No.	210–856–9

3. Synonyms, common and trade names

1-Chlor-2-propin
3-Chloro-1-propyne
3-Chlor-1-propin
Propargylchlorid

4.   Structural and molecular formulae

4.1   Structural formula $HC \equiv C\text{--}CH_2Cl$

4.2   Molecular formula C_3H_3Cl

5.   Physical and chemical properties

5.1   Molecular mass, g/mol   74.51

5.2   Melting point, °C   –78 (Budavari et al., 1989)

5.3   Boiling point, °C   57 (BASF, 1988)

5.4   Vapour pressure, hPa   245 (at 20 °C) (BASF, 1988)

5.5   Density, g/cm³   1.03 (at 20 °C)
(Budavari et al., 1989)

5.6   Solubility in water   Insoluble (Budavari et al., 1989)

5.7   Solubility in organic solvents   Insoluble in glycerine; miscible with benzene, carbon tetrachloride, ethanol, ethylene glycol, ether, ethyl acetate (Budavari et al., 1989)

5.8   Solubility in fat   No information available

5.9   pH-value   No information available

5.10 Conversion factor   1 ml/m³ (ppm) $\triangleq$ 3.04 mg/m³
1 mg/m³ $\triangleq$ 0.33 ml/m³ (ppm)
(at 1013 hPa and 25 °C)

6.   Uses

In the manufacture of corrosion inhibitors and polishing agents for galvanization (BASF, 1988).

7. **Experimental results**

7.1 Toxicokinetics and metabolism

No information available.

7.2 Acute and subacute toxicity

The approximate LD_{50} in the rat after oral administration was ca. 165 mg/kg body weight (no further details; BASF, 1987).

In an inhalation hazard test (at 25 °C) 6/6 and 1/6 rats died after exposure periods of 5 minutes and 1 minute, respectively, while 6/6 rats survived an exposure period of 30 seconds (calculated concentration 725 g/m^3, equivalent to about 240,000 ppm; BASF, 1987).

The following mortality figures were determined in further preliminary studies on the acute inhalation toxicity of propargyl chloride (see Table 1).

Table 1. Acute inhalation toxicity of propargyl chloride

Species	Nominal concentration	Duration of exposure	Mortality (dead/exposed)
Mouse	3000 mg/m^3 ($\triangleq$1000 ppm)	2 hours	6/10
Rat	3000 mg/m^3 ($\triangleq$1000 ppm)	2 hours	0/4
Guinea-pig	3000 mg/m^3 ($\triangleq$1000 ppm)	2 hours	0/1
Rabbit	3000 mg/m^3 ($\triangleq$1000 ppm)	2 hours	0/1
Cat	3000 mg/m^3 ($\triangleq$1000 ppm)	2 hours	1/1
Mouse	9000 mg/m^3 ($\triangleq$3000 ppm)	1 hour	2/10
Rat	9000 mg/m^3 ($\triangleq$3000 ppm)	1 hour	2/4
Guinea-pig	9000 mg/m^3 ($\triangleq$3000 ppm)	1 hour	1/1
Cat	9000 mg/m^3 ($\triangleq$3000 ppm)	1 hour	1/1

Post-mortem examination of the animals that died revealed pulmonary oedema or secondary bronchopneumonia (BASF, 1987).

In early studies, 10 mice, 4 rats, 1 guinea-pig, 1 rabbit and 1 cat were exposed to a nominal concentration of 300 mg/m^3, equivalent to 100 ppm for 3 hours/day on 5 consecutive days. All of the animals survived without showing specific signs of toxicity (BASF, 1987).

The sensory irritant effect of propargyl chloride (70% in toluene) was investigated in the "Alarie test". Groups of 4 male CD-1 mice (24 to 30 g) were acutely exposed to 0, 0.033, 0.323, 1.089 or 4.044 mg propargyl chloride/l air for 30 minutes. The respiratory frequency of the mice was recorded by means of whole-body plethysmography and a differentiation was made between an early (3 to 10 minutes) and a late phase (23 to 30 minutes). The exposure period of the mice at the highest concentration was extended to 60 minutes in order to detect possible damage to the lower respiratory tract (lung irritation). The criterion used was the RD$_{50}$, the concentration reducing the respiratory rate by 50%. From the percentage reductions at the various concentrations, this was calculated to be 2.0885 mg/l for the early phase, and 1.0303 mg/l for the late phase (literature value for toluene 12.68 mg/l). Immediately after the end of the study, the mice at the highest concentration (4.044 mg/l), showed increased tear secretion, which was reversible within 24 hours. Mice in the other concentration groups showed no further signs of toxicity other than depression of respiratory frequency. The reduction in respiratory rate was reversible in most of the test groups during the 10-minute observation phase, but the initial level was not reached again within this period in the 1.089 and 4.044 mg/l groups. The absolute and relative lungs weights were increased in the top dose group, in which exposure was extended to 60 minutes, which was seen as a sign of irritation of the lungs. It was concluded from the results of this study that propargyl chloride irritates the upper airways, with the contribution of toluene being negligible. Moreover, propargyl chloride causes mild irritation of the lungs (HRC, 1992).

In rats, the application of 2 ml concentrated propargyl chloride/rat to the shaved abdominal skin led to the death of all exposed animals after 10 minutes. Application of the same volume of a 10% solution in polyethylene glycol 400 for 4 hours was lethal to 2 of 6 exposed rats (BASF, 1987).

7.3   Skin and mucous membrane effects

Alcoholic solutions of propargyl chloride at concentrations of up to 1% did not cause skin irritation after application to the back of the rabbit for 20 hours under occlusive cover. A 10% solution led to "marked skin irritation", while concentrations of 20% or more gave rise to "severe necrotizing inflammation" and in some cases resulted in extravasation of blood and scar formation (no further information; BASF, 1987).

A 1% solution of propargyl chloride caused "mild inflammation that was reversible after 24 hours" when instilled into the rabbit eye. Higher concentrations were not tested (no further information; BASF, 1987).

7.4   Sensitization

No information available.

7.5   Subchronic and chronic toxicity

No information available.

7.6   Genotoxicity

7.6.1 In vitro

The genotoxicity of propargyl chloride (purity not specified) dissolved in DMSO was studied in the Salmonella/microsome test in strains TA 100, TA 98, TA 1535 and TA 1537. The concentrations used were 20 to 5000 µg/plate (up to 10,000 g/plate in TA 1535). The study was carried out with and without metabolic activation (S9-mix from Aroclor 1254-induced rat liver). Mutagenic activity occurred dose-dependently in strain TA 1535 at a concentration of 2500 µg/plate and above in two independent test series. The revertant count was at most 2.3-fold higher than the control value in the study with metabolic activation and 8.1-fold higher (at a concentration of 10,000 µg/plate) in the study without metabolic activation. There was a slight dose-dependent increase in the revertant count at 2500 µg/plate and above in strain TA 100 with and without metabolic activation and in strain TA 98 only without metabolic activation. At this concentration, the value of the revertant count was 1.5 to 2.6-fold above that of the control. No increase in the revertant count was seen in strain TA 1537. Because the slight mutagenic activity of the propargyl chloride used only occurred at high dose levels, the authors considered that this might have been due to the presence of contaminants (BASF, 1984).

Propargyl chloride (68.0% in toluene) was tested for mutagenic activity in a further Salmonella/microsome test in strains TA 1535, TA 1537, TA 1538, TA 98 and TA 100 with and without metabolic activation (S9-mix from Aroclor 1254-induced rat liver). The concentrations in two independent studies were 10, 33.3, 100, 333.3, 666.6 and 1000 µg/plate without S9-mix and 33.3, 100, 333.3, 1000 and 5000 µg/plate with S9-mix. Doses of 100 and 333.3 µg/plate (without activation) and 5000 µg/plate (with activation) were cytotoxic in TA 1537, similarly 5000 µg/plate was cytotoxic in TA 1538 (with activation). There was a concentration-dependent increase in the revertant count in strains TA 1535 and TA 100 with and without metabolic activation from 1000 and 100 µg/plate, respectively and the revertants were increased in TA 98 with and without metabolic activation (at 5000 µg/plate and from 100 µg/plate, respectively). Propargyl chloride was thus also shown to be mutagenic in this study (CCR, 1990).

7.6.2 In vivo

Propargyl chloride (ca. 70% in toluene, emulsified in olive oil) was studied for mutagenic activity in vivo in the micronucleus test in NMRI mice (average weight 27.5 g). At the same time, it was tested for possible activity as a spindle poison. The study was carried out in accordance with OECD guideline no. 474 and the EEC directive 84/449, B12. The maximum tolerated dose was ascertained in a preliminary study and was used as the highest dose. Groups of 5 males and 5 females were given single doses of 300, 100 and 33 mg/kg body weight by gavage. After 16, 24 and 48 hours (300 mg/kg) or after 24 hours (100 and 33 mg/kg), examination of the bone marrow of the femur took place with evaluation of 1000 polychromatic erythrocytes/mouse. A dose-related increase in toxicity was seen and 1 mouse given 300 mg/kg died on the following day. Examination of the bone marrow revealed no increase in the number of polychromatic erythrocytes with small or large micronuclei. Propargyl chloride therefore showed no clastogenic activity in this study and was not a spindle poison (BASF, 1992).

7.7   Carcinogenicity

No information available.

7.8   Reproductive toxicity

No information available.

7.9   Effects on the immune system

No information available.

7.10 Neurotoxicity

No information available.

7.11 Other effects

No information available.

8.   Experience in humans

No information available.

9.   Threshold limit values

No information available.

References

BASF AG, Abteilung Toxikologie
Report on the study of propargylchloride in the Ames test
Unpublished report (1984)

BASF AG, Gewerbehygienisch-pharmakologisches Institut
Unpublished studies
Written communication to BG Chemie (1987)

BASF AG, Abteilung DUS
Written communication to BG Chemie (1988)

BASF AG, Abteilung Toxikologie
Cytogenetic study in vivo of Propargylchlorid dest. ca. 70% in Tolu-
ol in mice, micronucleus test, single oral administration
Unpublished report project no. 26MO741/904505 (1992)

Budavari, S., O'Neil, M.J., Smith, A., Heckelman, P.E. (eds.)
The Merck index
11th ed., p. 1241
Merck & Co., Inc., Rahway, N.J., USA (1989)

CCR (Cytotest Cell Research GmbH & Co. KG, Rodorf)
Salmonella typhimurium reverse mutation assay with propargyl-
chloride (BG-no. 115)
Unpublished report CCR project 139601 (1990)
On behalf of BG Chemie

HRC (Huntingdon Research Centre Ltd., England)
Assessment of the respiratory tract irritancy of propargyl chloride
BG-no. 115, CAS-No. 624–65–7 (70% in toluene) in the mouse
Unpublished report BGH 27/911481 (1992)
On behalf of BG Chemie.

o-Chlorobenzotrichloride

This Toxicological Evaluation replaces a previously published version in volume 3

1. Summary and assessment

o-Chlorobenzotrichloride is of low toxicity to the rat on acute exposure (LD_{50} rat oral 1380 and 2390 mg/kg body weight; LD_{50} rat dermal >2000 mg/kg body weight). In an inhalation hazard test (7-hour exposure), o-chlorobenzotrichloride is tolerated by rats without signs of toxicity.

The chemical is irritating to the skin and eye of the rabbit. o-Chlorobenzotrichloride does not induce sensitization in guinea-pig skin in the maximization test.

o-Chlorobenzotrichloride induces mutations in the Salmonella/microsome test in the *Salmonella typhimurium* strains TA 98 and TA 100, in the presence but not in the absence of metabolic activation. It is not mutagenic in *Escherichia coli* either with or without S9-mix. No mutagenic activity has been found in the HPRT test with or without the addition of a metabolizing system. The substance is clastogenic in mice in the micronucleus test.

Tumour formation has been observed on the skin (1 squamous epithelioma), in the lungs (4 adenocarcinomas) and in the forestomach (1 carcinoma) with metastases in the liver, after dermal application to the mouse in a long-term study. Carcinogenic potential cannot be excluded for this chemical on the basis of these results.

2. Name of substance

2.1	Usual name	o-Chlorobenzotrichloride
2.2	IUPAC-name	$\alpha,\alpha,\alpha,2$-Tetrachlorotoluene

2.3	CAS-No.	2136–89–2
2.4	EINECS-No.	218–377–7

3. Synonyms, common and trade names

2-Chloro-1-trichloromethyl-
benzene
o-Chlorbenzotrichlorid
α,α,α-Trichloro-2- chlorotoluene
$\alpha,\alpha,\alpha,2$-Tetrachlortoluol

4. Structural and molecular formulae

4.1   Structural formula

4.2   Molecular formula $C_7H_4Cl_4$

5. Physical and chemical properties

5.1   Molecular mass, g/mol   229.9

5.2   Melting point, °C   30 (Weast, 1981/82)

5.3   Boiling point, °C   264.3 (Weast, 1981/82)

5.4   Vapour pressure, hPa   0.1 (at 20 °C) (Bayer, 1989)

5.5   Density, g/cm³   1.518 (at 20 °C)
(Weast, 1981/82)
1.458 (at 50 °C) (Bayer, 1989)

5.6   Solubility in water   Dissolves with difficulty
(Bayer, 1989)

5.7   Solubility in organic solvents   Soluble in ether, acetone
(Weast, 1981/82)

5.8 Solubility in fat	No information available
5.9 pH-value	Acid (Bayer, 1989)
5.10 Conversion factor	1 ml/m^3 (ppm) $\triangleq$ 9.38 mg/m^3
	1 mg/m^3 $\triangleq$ 0.11 ml/m^3 (ppm)
	(at 1013 hPa and 25 °C)

6.   Uses

Intermediate in the production of dyestuffs (Hoechst, 1991 b).

7.   Experimental results

7.1   Toxicokinetics and metabolism

No information available.

7.2   Acute and subacute toxicity

o-Chlorobenzotrichloride (97% pure) is of low acute toxicity. The oral LD$_{50}$ in rats was 1380 mg/kg body weight (1270 mg/kg body weight in males, 1490 mg/kg body weight in females). Deaths occurred up to 17 days after exposure. In addition to non-specific symptoms, signs of toxicity included impaired coordination and breathing, tremors, partly closed or closed eyelids and the production of clear to blood-coloured tears. At the high dose levels, the rats also showed impaired reflexes and narcosis. Bloody encrustation of the muzzles and the edges of the eyelids occurred independent of dose. The signs of toxicity were reversible in the surviving animals from day 13 (Hoechst, 1987 d).

In a further acute oral toxicity study in SPF-Wistar rats, an LD$_{50}$ value of 1.58 ml/kg body weight (equivalent to 2390 mg/kg) was reported. Single doses of 1.0 to 2.5 ml/kg body weight resulted in a deterioration in general condition as well as trembling spasms, weight loss, narcosis, impaired reflexes, ruffled fur, paralysis of the hind limbs, lacrimation, bloody eyes and lying on the stomach or side. The effects occurred 1 hour after administration, were not very pronounced, and persisted until the end of the 14-day observation period. Deaths occurred from day 3 to day 9 of the study. A dose of 0.5 ml/kg body weight was tolerated without signs of toxicity (Bayer, 1980 b).

In an inhalation hazard test in male and female Wistar rats, 200 l air/hour were passed through about 50 g of the test substance at 20 °C. The vapour-enriched air was then inhaled by 5 male and 5 female rats for 7 hours in a 10 l glass chamber. The observation period was 14 days. All of the rats tolerated the 7-hour exposure without symptoms. No deaths occurred. When the rats were dissected at the end of the observation period, macroscopic examination revealed that the organs were physiologically normal (Bayer, 1980 a).

No increase in mortality was observed in male and female Sprague-Dawley rats (weight 206 to 320 g, 8 to 10 weeks old) during a 14-day observation period following dermal application of o-chlorobenzotrichloride under occlusive cover for 24 hours at doses of 1000 and 2000 mg/kg body weight. After 4 hours, a red nasal secretion was seen in 2 of 5 males, but this was no longer evident after 24 hours.

Macroscopically, pale foci in the spleen and dilation of the kidney pelvis were observed in 2 males exposed to 2000 mg/kg. The dermal LD_{50} for o-chlorobenzotrichloride was consequently >2000 mg/kg body weight (IRI, 1992).

7.3   Skin and mucous membrane effects

The occlusive application of 500 mg o-chlorobenzotrichloride in the form of a paste to the clipped flank skin of New Zealand white rabbits (treated area 2.5×2.5 cm) for 24 hours led to clear skin irritation, which was evident immediately after removal of the patch and 72 hours after the start of the study. After 24 hours the skin had a grey-white coloration, and after 72 hours it had white to yellow blisters and marks. In addition, surface scurf and thickened, dry, flaking skin were observed. The substance was evaluated as severely irritating to the skin (FhG, 1980 a).

In a study in New Zealand white rabbits (1 or 2 per exposure period), 0.5 g of the test sample was applied to a cellulose patch (2.5 x

2.5 cm) which was fixed to the hairless skin of one ear with adhesive tape for 1, 2, 4, 8 or 24 hours. The observation period lasted 7 days. Skin exposed for 1 or 2 hours was physiologically normal when the dressings were removed. Rabbits that were exposed for longer periods had moderate to very marked skin reddening up until the end of the observation period and, when exposure was for 8 hours or more, slight to moderate swelling of the skin. The study revealed a severe irritant effect on the skin (Bayer, 1980 a).

In a further study, 0.5 ml of the chemical (97% pure) was applied semi-occlusively to the clipped backs of rabbits (New Zealand white) for 4 hours. 1 hour after the patch was removed, all of the animals had barely perceptible erythema and 2 of the rabbits had very slight oedema. 24 to 72 hours after the patch was removed the rabbits had barely perceptible to moderate erythema and barely perceptible to slight oedema. 7 days after application, a clear localized erythema was present in all animals and slight oedema was seen in 2 rabbits. In addition, 72 hours after removal of the patch, the skin was dry and chapped in places and had small areas of brown discoloration. 7 days after application the skin was scurfy and had parchment-like cracks, and fine flakes were becoming detached. All of the effects were reversed 14 days after application. In this study, the substance proved to be irritating to the skin (Hoechst, 1987 b).

In a study designed to test the irritant effect of o-chlorobenzotrichloride on the eye, 100 µl of the substance was applied as a paste to the conjunctival sac of the lower right eyelid of 6 male rabbits. The untreated left eyes served as controls in each case. The results were evaluated 24, 48 and 72 hours and 8 days after the beginning of the study in accordance with the scheme devised by Draize (1959). Neither the cornea nor the iris were damaged in any of the animals. 2 of the 6 rabbits had slight reddening of the conjunctiva (grade 1) after 24 hours and 1 animal showed a slight swelling (grade 1). The effects were reversible after 8 days. The substance was evaluated as not irritating to the rabbit eye (FhG, 1980 b).

o-Chlorobenzotrichloride was instilled into the conjunctival sac of one eye in each of 2 rabbits at a dose of ca. 50 mg/rabbit. The untreated eyes served as controls. The observation period lasted 7 days. Moderate reddening and slight swelling of the conjunctiva were observed. 6 days after exposure, the treated eyes were once again physiologically normal. In this study o-chlorobenzotrichloride was evaluated as moderately irritating to the eye of the rabbit (Bayer, 1980 a).

In a further study, 0.1 ml of the test substance (97% pure) was instilled into the left conjunctival sac of each of 3 New Zealand white rabbits. The untreated right eyes served as controls. The eyes were evaluated 1, 24, 48 and 72 hours after instillation. The cornea and the iris were undamaged. The conjunctiva was markedly reddened after 1 to 24 hours and slightly red after 48 hours. Slight to marked swelling of the conjunctiva was present at 1 and 24 hours. In addition, a clear, colourless secretion was observed after 1 hour. After 48 hours, marked hyperaemia of some of the conjunctival blood vessels was seen, which was still detectable in one of the rabbits after 72 hours. This rabbit showed no irritant effects after 7 days (Hoechst, 1987 c). The substance caused irritation that was reversible within 7 days.

7.4   Sensitization

The sensitizing potential of o-chlorobenzotrichloride was studied in the Magnusson-Kligman maximization test in 20 female albino guinea-pigs (Dunkin-Hartley). For induction, the test substance was initially administered intradermally (10% in paraffin oil) and then applied dermally 7 days later (25% in paraffin oil). The challenge was carried out dermally (25% in paraffin oil), 3 weeks after the beginning of the induction treatment. No skin reactions were observed, and o-chlorobenzotrichloride was thus not shown to induce sensitization in this study protocol (IRI, 1991).

7.5   Subchronic and chronic toxicity

No information available.

7.6   Genotoxicity

7.6.1 In vitro

o-Chlorobenzotrichloride (97% pure) was tested in the Ames test in the *Salmonella typhimurium* strains TA 98, TA 100, TA 1535,

TA 1537 and TA 1538 and in *Escherichia coli* WP2uvrA with and without a metabolic activation system (S9-mix from Aroclor 1254-induced rat liver). The doses used were between 0.16 and 5000 µg/plate. o-Chlorobenzotrichloride did not induce mutations in any of the strains used in the absence of metabolic activation. On addition of S9-mix, an increase in the revertant count was found in strains TA 98 and TA 100. o-Chlorobenzotrichloride was mutagenic in this test system in the presence of a metabolic activation system (Hoechst, 1987 a).

o-Chlorobenzotrichloride (98.6% pure) was studied in the HPRT test in Chinese hamster V79 cells with and without metabolic activation (S9-mix from the livers of Aroclor 1254-treated male Wistar rats). Concentrations of from 2 to 40 µg/ml without S9-mix and 4 to 40 µg/ml with S9-mix were tested in two independent experiments. The highest concentration was cytotoxic without metabolic activation. With metabolic activation, no toxic effects were observed at concentrations of up to 40 µg/ml. Under these study conditions, o-chlorobenzotrichloride was not mutagenic at the HPRT locus (CCR, 1991 a).

7.6.2 In vivo

In a micronucleus test, o-chlorobenzotrichloride (98.6% pure) was administered to male and female NMRI mice (weight ca. 30 g) as a single oral dose of 100, 333 or 1000 (maximum tolerated dose) mg/kg body weight in polyethylene glycol 400. Mice that were given cyclophosphamide (40 mg/kg body weight) served as positive controls. Compared with the negative controls, which received only the vehicle (polyethylene glycol 400), a dose-dependent increase in the occurrence of micronuclei in the bone marrow cells of treated rats was observed 24 hours after administration. A statistically significant increase was found at the highest dose level after 24, 48 and 72 hours. o-Chlorobenzotrichloride was thus mutagenic under these study conditions (CCR, 1991 b).

7.7   Carcinogenicity

A preparation of o-chlorobenzotrichloride in benzene was applied to the clipped dorsal skin (shoulder blade) of 20 female mice (strain

ICR-SLC, 7 weeks old) at a dose of 5 l/mouse, twice a week for 30 weeks. The mice were then observed up to 18 months. A control group (20 mice) was treated with 25 l benzene. The results of this long term study are presented in the following table.

Table 1. Tumour spectrum in female mice after dermal application of o-chlorobenzotrichloride for 30 weeks

Substance	Number of mice	Dose per mouse	Age at end of study (days)	Number of animals examined	Tumours skin	lungs	gastro-intestinal tract
Benzene (control)	20	25 µl	551	17	–	–	–
o-Chloro-benzotri-chloride	20	5 µl	553	18	1[a]	4[b]	1[c]

[a] Squamous epithelioma
[b] Adenocarcinomas
[c] Forestomach carcinoma with formation of metastases in the liver

The authors interpreted the results as showing a significant increase in the tumour rate, even though it was clearly lower than that seen with p-chlorobenzotrichloride or benzotrichloride (Matsushita et al., 1977). However, only one sex, one dose group and 20 animals were tested in this study. There were no details of non-neoplastic effects and no precise description of the method was given. No historical tumour incidences were provided. Despite the weaknesses of the study method, carcinogenic potential cannot be excluded based on these data.

7.8   Reproductive toxicity

No information available.

7.9   Effects on the immune system

No information available.

7.10 Neurotoxicity

No information available.

7.11 Other effects

No information available.

8. Experience in humans

No information available.

9. Threshold limit values

No information available.

References

Bayer AG, Institut für Toxikologie
o-Chlorbenzotrichlorid - Gewerbetoxikologische Untersuchungen
Unpublished report no. 9064 (1980 a)

Bayer AG, Institut für Toxikologie
o-Chlorbenzotrichlorid destl. - Untersuchungen zur akuten oralen
Toxizität an männlichen und weiblichen Wistar-Ratten
Unpublished report (1980 b)

Bayer AG, Geschäftsbereich Organica
Safety data sheet, o-Chlorbenzotrichlorid destilliert (1989)

CCR (Cytotest Cell Research GmbH & Co. KG)
Gene mutation assay in Chinese hamster V79 cells in vitro with
o-Chlorbenzotrichlorid
Unpublished report, CCR project 174824 (1991 a)
On behalf of BG Chemie

CCR (Cytotest Cell Research GmbH & Co. KG)
Micronucleus assay in bone marrow cells of the mouse with
o-Chlorbenzotrichlorid
Unpublished report, CCR project 174813 (1991 b)
On behalf of BG Chemie

FhG (Fraunhofer-Institut für Toxikologie und Aerosolforschung)
Bericht über die Prüfung von o-Chlorbenzotrichlorid auf primäre
Hautreizwirkung
Unpublished report (1980 a)
On behalf of Bayer AG

FhG (Fraunhofer-Institut für Toxikologie und Aerosolforschung)
Bericht über die Prüfung von o-Chlorbenzotrichlorid auf Schleim-
hautreizwirkung
Unpublished report (1980 b)
On behalf of Bayer AG

Hoechst AG, Pharma Forschung Toxikologie und Pathologie
o-Chlorbenzotrichlorid TTR - Study of the mutagenic potential in
strains of *Salmonella typhimurium* (Ames test) and *Escherichia coli*
Unpublished report no. 87.0528 (1987 a)

Hoechst AG, Pharma Forschung Toxikologie und Pathologie
o-Chlorbenzotrichlorid TTR - Prüfung auf Hautreizung am Kaninchen
Unpublished report no. 87.0479 (1987 b)

Hoechst AG, Pharma Forschung Toxikologie und Pathologie
o-Chlorbenzotrichlorid TTR - Prüfung auf Augenreizung am Kaninchen
Unpublished report no. 87.0565 (1987 c)

Hoechst AG, Pharma Forschung Toxikologie und Pathologie
o-Chlorbenzotrichlorid TTR - Prüfung der akuten oralen Toxizität
an der männlichen und weiblichen Wistar-Ratte
Unpublished report no. 87.0566 (1987 d)

Hoechst AG
Communication to BG Chemie of 16.01.1991 a

Hoechst AG
Communication to BG Chemie of November 1991 b

IRI (Inveresk Research International Ltd.)
o-Chlorobenzotrichloride: Magnusson-Kligman maximisation test in guinea pigs
Unpublished report no. 8204 (1991)
On behalf of BG Chemie

IRI (Inveresk Research International Ltd.)
o-Chlorobenzotrichloride: acute dermal toxicity (LD_{50}) test in rats
Unpublished report no. 8286 (1992)
On behalf of BG Chemie

Matsushita, H., Fukuda, K., Takemoto, K.
Carcinogenicity of benzotrichloride and its related compounds
Unpublished report, National Institute of Industrial Health, Ministry of Labour, Japan (1977)

Weast, R.C. (ed.)
CRC Handbook of chemistry and physics
62nd ed., p. C-166
CRC Press, Boca Raton, Florida (1981/1982)

2-Methoxy-2,3-dihydropyran

1. Summary and assessment

2-Methoxy-2,3-dihydropyran is of low to moderate toxicity on acute oral administration (LD_{50} rat oral, between 1410 and 3740 mg/kg body weight; LD_{50} rabbit oral ca. 250 mg/kg body weight). The LD_{50} in the rabbit after dermal application is 4920 mg/kg body weight. Signs of toxicity observed after acute exposure include lacrimation, apathy, ataxia, reduced muscle tone, reduced reaction or no reaction to external stimuli, dyspnoea and reduced body weight gain. The acute toxicity on inhalation exposure (LC_{50}) in a dynamic whole-body inhalation chamber is low in the rat, and is given as > 6100 mg/m^3 air after exposure for 4 hours, with effects seen including nasal secretion, dyspnoea and ataxia.

Pure 2-methoxy-2,3-dihydropyran has a clear irritant effect on the skin after occlusive exposure for 4 hours. A more severe irritant effect, which may extend to surface scar formation, is observed on longer exposure, with contaminants possibly increasing the irritant effect. No studies carried out in accordance with EC criteria are available, but an irritant effect as defined by the EC criteria should be assumed. Pure 2-methoxy-2,3-dihydropyran (> 99%) is marginally irritating to the eye of the rabbit which, in accordance with EC criteria, is evaluated as not irritating. In contrast, effects that were found with a substance that was not analytically characterized, were not reversible after a comparable observation period. Analytically uncharacterized preparations of 2-methoxy-2,3-dihydropyran should therefore be considered as irritating. Different grades of purity may possibly be responsible for the different results observed, as the study methods appear to be identical based on their description.

The ability of technical grade 2-methoxy-2,3-dihydropyran to induce skin sensitization cannot be excluded based on the one available study in guinea-pigs.

In the Salmonella/microsome test in 4 strains, 2-methoxy-2,3-

dihydropyran is only mutagenic in TA 100 with metabolic activation. The lowest mutagenic concentration is given as 500 µg/plate. Without metabolic activation, there are no indications of mutagenic activity in prokaryotes. The substance does not induce chromosome damage in the micronucleus test in the mouse in vivo.

2.   Name of substance

2.1	Usual name	2-Methoxy-2,3-dihydropyran
2.2	IUPAC-name	2-Methoxy-2,3-dihydro-4H-pyran
2.3	CAS-No.	4454–05–1
2.4	EINECS-No.	224–698–3

3.   Synonyms, common and trade names

2,3-Dihydro-2-methoxy-4H-pyran
3,4-Dihydro-2-methoxy-2H-pyran
3,4-Dihydro-2-methoxypyran
Methoxydihydropyran
2-Methoxy-3,4-dihydro-2H-pyran
2-Methoxy-3,4-dihydropyran

4.   Structural and molecular formulae

4.1   Structural formula

4.2   Molecular formula $C_6H_{10}O_2$

5.   Physical and chemical properties

5.1   Molecular mass, g/mol   114.15

5.2	Melting point, °C	< -60 (BASF, 1985)
5.3	Boiling point, °C	ca. 126 (BASF, 1985) 128.5 (Dodd et al., 1988)
5.4	Vapour pressure, hPa	ca. 10 (at 20 °C) (Dodd et al., 1988) 13 (at 20 °C) (BASF, 1991) 66 (at 50 °C) (BASF, 1991)
5.5	Density, g/cm^3	1 (at 20 °C) (BASF, 1985) 1.003 (at 20 °C) (Dodd et al., 1988)
5.6	Solubility in water	16 g/l (at 25 °C) (BASF, 1991) 13 g/l (at 30 °C) (BASF, 1985)
5.7	Solubility in organic solvents	Unlimited solubility in methanol (at 20 °C) (BASF, 1985)
5.8	Solubility in fat	Partition coefficient n-octanol/ water, log P_{ow}: 1.33 (BASF, 1991)
5.9	pH-value	-
5.10	Conversion factor	1 ml/m^3 (ppm) $\triangleq$ 4.67 mg/m^3 1 mg/m^3 $\triangleq$ 0.21 ml/m^3 (ppm) (at 1013 hPa and 25 °C)

6.   Uses

Intermediate in the manufacture of biocides, tannins, pharmaceuticals and pesticides (BASF, 1993 a).

7.   Experimental results

7.1   Toxicokinetics and metabolism

No information available.

7.2   Acute and subacute toxicity

The LD_{50} values presented in Table 1 have been determined after single oral, intraperitoneal and dermal exposures.

Table 1. Data on the acute toxicity of 2-methoxy-2,3-dihydropyran

Species	Sex	Route	LD$_{50}$ value (mg/kg b.w., 95% confidence limits	References
Rat	female	oral	1410	Dodd et al., 1988
Rat	male	oral	1870	Dodd et al., 1988
Rat	female	oral	3600 (3280–3950)	INBIFO, 1977 a
Rat	male/female	oral	3740 (3530–3970)	INBIFO, 1977 a
Rat	male	oral	3890 (3660–4140)	INBIFO, 1977 a
Rabbit	no data	oral	ca. 250	BASF, 1975 a*
Rabbit	male/ female	dermal	> 200	INBIFO, 1977 c; BASF, 1980
Rabbit	male/ female	dermal	4920	Dodd et al., 1988
Mouse	male	i.p.	519 (370–729)	INBIFO, 1977 b
Mouse	male/ female	i.p.	525 (417–661)	INBIFO, 1977 b; BASF, 1980
Mouse	female	i.p.	575 (467–706)	INBIFO, 1977 b

* Both pure and crude 2-methoxy-2,3-dihydropyran tested; approximate LD$_{50}$
b.w. body weight

The following signs of toxicity were seen in mice, rats and rabbits during studies on the acute toxicity of 2-methoxy-2,3-dihydropyran after oral, intraperitoneal and dermal administration: apathy, ataxia, partly closed eyes, lacrimation, ruffled fur, dyspnoea, reduced muscle tone, disturbances in coordination, tremors, a narcosis-like state and lying on the stomach or side. Body weight gain was inhibited. Post-mortems on the animals that died spontaneously showed, in mice, rats and rabbits, macroscopic effects on the liver (marbling, grey-white foci in the hilus region, doughy consistency), heart (dilation, occasional hydrothorax), lungs (red discoloration), stomach (oedema of the stomach wall, gastric mucosa covered with eschar), gut (reddened, occasionally with subserous bleeding) and kidneys (loss of zonal organization). Post-mortems on the animals surviving the 14-day observation period revealed no treatment-related effects on the organs (BASF, 1975 a; 1980; INBIFO, 1977 a, b, c; Dodd et al., 1988).

A single oral dose of "pure" or "crude" 2-methoxy-2,3-dihydropyran given to rabbits (3/group) by gavage at a dose of 0.2 ml/kg body weight as a 2% solution or emulsion respectively (equivalent to 200 mg/kg) had no effect on liver function. The parameters studied were alanine aminotransferase activity and bromsulphthalein retention. The parameters were determined immediately before administration of the substance and 8 and 14 days after the start of the study. 1 of the 3 rabbits used per group died in each case. Post-mortems of the rabbits that died revealed acute dilation of the heart on both sides. The musculature of the left chamber of the heart was pale and the mucosa of the stomach diffusely reddened. No treatment-related pathological effects on the organs were observed in the rabbits that survived the 14-day observation period (BASF, 1975 a).

Sprague-Dawley rats (10 male and 10 female, weight 185±15 g) were exposed to an average concentration of 6.1 mg 2-methoxy-2,3-dihydropyran/l air (analytical concentration) in a dynamic whole-body inhalation chamber for 4 hours. The control animals (10/sex) inhaled air. The observation period was 14 days. The test substance was ca. 95% pure (no information on contaminants). None of the animals in the treated or control groups died during the 4-hour exposure or the 14-day observation period. The 4-hour LC_{50} was thus > 6.1 mg/l air. Body weight gain was unimpaired in the treated group. Signs of toxicity seen in the treated rats included watery to red nasal secretion and subsequently red, gummed-up eyes and noses, closed eyes, jerky breathing, reeling gait, and ruffled, matted fur. From day 8 of the study, the animals appeared normal. Post-mortems revealed no treatment-related effects on the organs (BASF, 1979).

Exposure of male and female Sprague-Dawley rats to an atmosphere that was approaching saturation with 2-methoxy-2,3-dihydropyran (99.1 % pure; other components 0.64 % dimers of the test substance, 0.037 % acrolein) for 1 hour in a static whole-body inhalation chamber caused 100 % mortality. The rats died either during exposure or within 24 hours after exposure. The test was repeated in an independent experiment. The analytical concentrations of 2-methoxy-2,3-dihydropyran were on average 8098±269 ppm and 8044±878 ppm in experiments 1 and 2, respectively (equivalent to ca. 38,000 mg/m^3 air). The acrolein concentration (only determined in experiment 2) was 240±41 ppm. 5 males and 5 females were used per experiment. The observation period was 14 days. Signs of toxicity observed included lacrimation, watery secretions from the nose and mouth, dyspnoea, reduced mobility, prostration and loss of righting reflexes. Autopsy revealed haemorrhagic lungs and clear fluid in the trachea and thoracic cavity. The authors suggested that the toxicity of 2-methoxy-2,3-dihydropyran could be due to contamination with acrolein. The LC$_{50}$ value for acrolein was 8.3 ppm in rats after a 4-hour exposure and 26 ppm after a 1-hour exposure (Ballantyne et al., 1989; Dodd et al., 1988).

Sprague-Dawley rats (5 male and 5 female) were exposed to an average analytically determined 2-methoxy-2,3-dihydropyran concentration of 7748 ±524 ppm (equivalent to ca. 36,000 mg/m^3 air) for 1 hour in a dynamic whole-body inhalation chamber. The test substance was 99.1 % pure (other components 0.64 % dimers of the test substance, 0.037 % acrolein). Acrolein could not be detected in the dynamic inhalation chamber (detection limit 1 ppm). The observation period was 14 days. No mortality occurred. During exposure, the rats showed lacrimation and hyperactivity. Immediately after exposure, watery secretions from the nose and mouth and reduced mobility were observed. No further signs of toxicity occurred. Body weight gain was normal. No macroscopically visible treatment-related effects on the organs were evident at autopsy (Ballantyne et al., 1989; Dodd et al., 1988).

In an inhalation hazard test, rats were exposed to an atmosphere enriched with 2-methoxy-2,3-dihydropyran vapour at 20 °C for 1 or 3 hours. To enrich the atmosphere, 200 l air/hour was passed through a 5 cm deep layer of 2-methoxy-2,3-dihydropyran (contaminants 4% dimeric acrolein and traces of acrolein and methylvinyl ether). None of the 12 exposed rats died after exposure for 1 hour, while after 3 hours, 10 of the 12 rats had died. Signs of toxicity that occurred during exposure included escape attempts, closed eyes, watery nasal secretions, salivation, loss of pain reflexes, jerky breathing, lying on the side, narcosis, ruffled fur, and, in 2 rats, a milky, clouded cornea. Post-mortems on the 10 rats that died during the study showed acute dilation of the right side, and acute congestive hyperaemia on both sides of the heart, while the lungs repeatedly showed moderate acute swelling, moderate congestion and moderate oedema. No treatment-related macroscopic effects on the organs were seen in the 2 rats dissected at the end of the 14-day observation period (BASF, 1980).

7.3   Skin and mucous membrane effects

The skin irritancy of "pure" and "crude" 2-methoxy-2,3-dihydropyran was tested on the clipped dorsal skin of white rabbits (2 in each case). No data on purity or other components present was provided. The treated skin was examined 24 hours and 8 days after a 20-hour occlusive exposure to a patch soaked with the neat test substance. After 24 hours, the 2-methoxy-2,3-dihydropyran designated as "pure" caused slight reddening of the skin in both rabbits, which was reversible by the end of the study (day 8). At this point, one of the rabbits showed slight flaking of the treated area. In contrast, "crude" 2-methoxy-2,3-dihydropyran had a clear irritant effect. Slight redness was seen in both rabbits and one of them also showed severe oedema formation 24 hours after removal of the patch. At the end of the study (day 8) severe scaling was observed in one of the rabbits, while the animal with oedema formation showed a slight, superficial necrosis. The "crude" substance was thus irritating to the skin (BASF, 1975 b).

In a Draize test, 6 rabbits were treated with 0.5 ml neat 2-methoxy-2,3-dihydropyran (purity unspecified), which was applied to the intact and scarified dorsal skin under occlusive conditions for 24 hours. No effects on the skin were observed 24 or 72 hours after application of the substance. However, 7 days after exposure, minor effects on the skin (induration, thickening) were described. Based on these effects, 2-methoxy-2,3-dihydropyran was evaluated as slightly irritating. The effects on intact and scarified skin did not differ (INBIFO, 1977 d; BASF, 1980).

In a further skin irritation study, 0.5 ml neat 2-methoxy-2,3-dihydropyran (99.105% pure, other components 0.643% dimers of the test substance, 0.037% acrolein) was applied occlusively to the clipped dorsal skin of 3 male and 3 female rabbits for 4 hours. The findings were evaluated 1 hour and 1, 2, 3 and 7 days after removal of the patch. Apart from on day 7, marked reddening and moderate to severe oedema were observed at all time points. The substance thus showed a clear irritant effect (Dodd et al., 1988).

To study eye irritancy, 0.1 ml neat 2-methoxy-2,3-dihydropyran (purity not specified) was instilled into the conjunctival sac of one eye of each of 6 rabbits. The eyes were examined 24, 48 and 72 hours and 7 days after treatment. In each case the untreated eye served as a control. The test substance led to extensive clouding of the cornea which became more severe with time, and which was not reversible in 3 of the rabbits by the end of the study (day 7). The reaction of the conjunctiva was also pronounced in all of the animals. After 72 hours grade 1 redness, grade 2 to 3 swelling, grade 2 secretion and signs of secondary infection (suppuration) were still present. No indication was given of the reversibility of these effects at the end of the study. The average irritation score (33.4 from a maximum 110) led to the substance being evaluated as "moderately irritating" by the authors (INBIFO, 1977 e; BASF, 1980).

In a further eye irritation study, 6 rabbits each had 0.1 ml neat 2-methoxy-2,3-dihydropyran (99.105% pure, other components 0.643% dimers of the test substance, 0.037% acrolein) instilled into the conjunctival sac of one eye. The findings were recorded 1 and 4 hours and 1, 2, 3 and 7 days after instillation. In each case the untreated eye served as a control. The substance had no effects on the cornea but caused iritis that was reversible within 24 hours. The similarly slight inflammation of the conjunctiva (redness, swelling maximum grade 1.3 and discharge grade 2 to 3) diminished during the first 3 days after instillation and was completely reversible within 7 days. The authors also studied the eye irritancy of smaller amounts (0.01 and 0.005 ml) of 2-methoxy-2,3-dihydropyran under comparable test conditions, where, as expected, a slight weakening of the effects was evident (Dodd et al., 1988).

7.4   Sensitization

20 female Pirbright white guinea-pigs and 6 female controls had the hair removed from the front of their left and right flanks (area ca. 25 cm^2). After 4 hours the left flank was defatted with ether, and a cotton wool pad that had been soaked with either "pure" or "crude" neat 2-methoxy-2,3-dihydropyran (10 animals/group) was used to paint 3 crosses on the skin, one behind the other. No details on the purity of the substances were provided. The application was repeated twice daily on the following days (5 times a week=10 paintings in all). Application of the pure test substance in undiluted form led to slight to severe flaking and slight oedema in the area of application in all 10 animals, while with the crude test substance scab formation occurred in the area of application in all 10 animals. After a treatment-free period of 11 days, during which the inflammatory effects on the left flank caused by the treatment cleared up, the hair was again removed from both flanks. 4 hours later, after defatting with ether, the right, previously untreated flank was painted once with either the pure or the crude test substance in undiluted form, and the skin reaction was recorded after 12 hours. The control groups (3 animals in each case) were treated in the same way with either the pure or the crude test substance. The pure test substance did not induce skin reactions in any of the 10 previously exposed animals, and only a yellowish-brown residue of the substance was found. The crude test substance caused mild, occasionally mottled redness in 6 out of 10 of the previously exposed animals and in addition, the application site showed red-brown residues of the test substance. There were no visible irritant effects in the 6 untreated controls after a single application to the right flank, only red-brown residues of the test substance. The challenge treatment was there-

fore repeated in the test substance-treated animals after a further 8 days. In this case, a single application of the neat crude test substance to the right flank caused indications of redness in all 10 animals, and residues of the test substance were present at the application site. The skin of the 3 untreated controls showed no inflammatory effects. The induction of skin sensitization by "crude" 2-methoxy-2,3-dihydropyran could not be excluded based on the results of this study (BASF, 1975 c).

7.5   Subchronic and chronic toxicity

No information available.

7.6   Genotoxicity

7.6.1 In vitro

2-Methoxy-2,3-dihydropyran (> 99.9% pure) was tested for mutagenicity in the *Salmonella typhimurium* strains TA 98, TA 100, TA 1535 and TA 1537 at concentrations of 20 to 5000 µg/plate in the standard plate test. 3 plates were used per concentration and strain. The study was carried out with and without metabolic activation (S9-mix from Aroclor 1254-induced rat liver). No mutagenic activity was observed in any of the test strains without metabolic activation. With metabolic activation, the revertant count was increased concentration-dependently in strain TA 100 only, from 500 µg/plate. The maximum increase seen was 3.4-fold over the control value (DMSO carrier) at the highest concentration (5000 µg/plate). In a second independent study protocol, the test substance was investigated in strain TA 100 at concentrations of from 100 to 7500 µg/plate (3 plates/concentration) with and without metabolic activation. Once again, there were no indications of mutagenic activity in strain TA 100 without metabolic activation. With metabolic activation, the revertant count was increased concentration-dependently from 2500 µg/plate, to a maximum of 4.3-fold over the control value at the highest concentration (BASF, 1989).

7.6.2 In vivo

The clastogenic activity of 2-methoxy-2,3-dihydropyran was investigated in a micronucleus test (in accordance with OECD guideline no. 474) in male and female NMRI mice (ca. 26 g). The substance (purity 99.9%) was dissolved in olive oil and administered once by gavage at doses of 1000, 500 and 250 mg/kg body weight (volume administered 10 ml/kg). Negative controls were given the solvent, positive controls 20 mg cyclophosphamide/kg body weight or 0.15 mg vincristine/kg body weight. Bone marrow preparations were made 16, 24 and 48 hours after treatment in the group given 1000 mg/kg, and 24 hours after administration in the other groups. The number of polychromatic erythrocytes evaluated was 1000/ mouse. A single oral dose of 2-methoxy-2,3-dihydropyran did not lead to an increase in micronucleus-containing polychromatic erythrocytes at any of the doses tested in this study. The ratio of polychromatic to normochromatic erythrocytes was in the range of control values in all dose groups. The substance was thus not shown to be clastogenic (BASF, 1993 b).

7.7   Carcinogenicity

No information available.

7.8   Reproductive toxicity

No information available.

7.9   Effects on the immune system

No information available.

7.10 Neurotoxicity

No information available.

7.11 Other effects

No information available.

8.   Experience in humans

No information available.

9.   Threshold limit values

No information available.

References

Ballantyne, B., Dodd, D.E., Pritts, I.M., Nachreiner, D.J., Fowler, E.H.
Acute vapour inhalation toxicity of acrolein and its influence as a
trace contaminant in 2-methoxy-3,4-dihydro-2H-pyran
Hum. Toxicol., 8, 229–235 (1989)

BASF AG, Gewerbehygiene und Toxikologie
Bericht über die Prüfung von Methoxidihydropyran, Methoxidihy-
dropyran roh auf etwaige leberschädigende Wirkung
Unpublished report, studies no. XXIII/78 and XXIII/114 (1975 a)

BASF AG, Gewerbehygiene und Toxikologie
Bericht über die Prüfung der Hautreizwirkung von Methoxidihy-
dropyran im Vergleich zu Methoxidihydropyran roh am Kaninchen
Unpublished report, studies no. XXIII/78 and XXIII/114 (1975 b)

BASF AG, Gewerbehygiene und Toxikologie
Bericht über die Prüfung von Methoxidihydropyran im Vergleich zu
Methoxidihydropyran, roh auf etwaige hautsensibilisierende Wir-
kung
Unpublished report, studies no. XXIII/78 and XXIII/114 (1975 c)

BASF AG, Gewerbehygiene und Toxikologie
Bericht über die Bestimmung der akuten Inhalationstoxizität LC_{50}
von Methoxidihydropyran als Dampf bei 4 stündiger Exposition an
Sprague-Dawley-Ratten, Substanz-Nr. 77/422
Unpublished report (1979)

BASF AG, Gewerbehygiene und Toxikologie
Bericht über die gewerbetoxikologische Grundprüfung, Substanz-
Nr. 77/422
Unpublished report (1980)

BASF AG
DIN-Safety data sheet 2-Methoxy-2,3-dihydro-4H-pyran (1985)

BASF AG, Abteilung Toxikologie
Report on the study of 2-methoxi-2,3-dihydro-4H-pyran in the
Ames test
Unpublished report, project no. 40M0968/884404 (1989)

BASF AG
AIDA-Grunddatensatz 3,4-Dihydro-2-methoxy-2H-pyran (1991)

BASF AG
Communication to BG Chemie of 19.03.1993 a

BASF AG, Abteilung Toxikologie
Cytogenetic study in vivo of 2-methoxy-2,3-dihydro-4H-pyran in
mice - micronucleus test, single oral administration
Unpublished report, project no. 26M0015/914321 (1993 b)

Dodd, D.E., Ballantyne, B., Myers, R.C., Pritts, I.M., Nachreiner,
D.J.
The acute toxicity of methoxydihydropyran
Vet. Hum. Toxicol., 30, 545–550 (1988)

INBIFO (Institut für biologische Forschung)
Methoxidihydropyran - Akute Toxizitätsbestimmung an der
männlichen und weiblichen Ratte, orale (intragastrale) Applikation
Unpublished report no. A 0135/1471 (1977 a)
On behalf of BASF AG

INBIFO (Institut für biologische Forschung)
Methoxidihydropyran - Akute Toxizitätsbestimmung an der
männlichen und weiblichen Maus, intraperitoneale Applikation

Unpublished report no. A 0135/1478 (1977 b)
On behalf of BASF AG

INBIFO (Institut für biologische Forschung)
Methoxidihydropyran - Prüfung auf systemische dermale Toxizität
am Kaninchen
Unpublished report no. A 0135/1511 (1977 c)
On behalf of BASF AG

INBIFO (Institut für biologische Forschung)
Methoxidihydropyran - Prüfung auf Reizwirkung an der Kaninchen-
haut
Unpublished report no. A 0135/1485 (1977 d)
On behalf of BASF AG

INBIFO (Institut für biologische Forschung)
Methoxidihydropyran - Prüfung auf Reizwirkung am Kaninchen-
auge
Unpublished report no. A 0135/1493 (1977 e)
On behalf of BASF AG

Ohara, T., Sato, T., Shimizu, N., Prescher, G., Schwind, H., Wei-
berg, O.
Acrolein and methacrolein
In: Ullmann's encyclopedia of industrial chemistry
5th ed., vol. A1, pp. 149–159
VCH Verlagsgesellschaft, Weinheim (1985)

p-Chlorobenzotrichloride

This Toxicological Evaluation replaces a previously published version in volume 3.

1.   Summary and assessment

After administration of a single oral dose of p-chlorobenzotrichloride, the substance is metabolized and eliminated within 4–6 days, mainly in the urine (77 to 87%) and to a lesser extent in the faeces (9 to 14%). The amount remaining in the tissues is about 4%. 4-Chlorohippuric acid appears as the main metabolite (>90%) in the urine. The metabolites 4-chlorobenzoic acid, 4-chlorohippuric acid and $\alpha,\alpha',4,4'$-tetrachlorostilbene have been detected in small amounts in the faeces.

p-Chlorobenzotrichloride is of moderate acute toxicity (LD_{50} rat oral 614 to 1350 mg/kg body weight; LD_{50} mouse oral 700 mg/kg body weight; LD_{50} rat dermal 1900 mg/kg body weight). On inhalation exposure, the substance is evidently more toxic (LC_{50} rat and mouse 125 mg/m^3; no data on duration of exposure).

After repeated oral administration of p-chlorobenzotrichloride to rats for 14 days, dose-dependent signs of toxicity such as reduced body weight gain and feed intake, body weight loss, gastrointestinal disturbances, general deterioration in condition, dehydration, impaired breathing, ataxia and spasms were evident at 25 to 300 mg/kg body weight. The 300 mg/kg body weight dose level was lethal to all of the rats, while 2 out of 12 died at 150 mg/kg body weight. No clinical signs of toxicity or treatment-related macroscopic effects were observed at 1.25 mg/kg body weight (no effect level). Inhalation of 4, 20 or 100 mg p-chlorobenzotrichloride/m^3 for 30 days was predominantly marked by clinical signs of irritation of the airways at the highest concentration. Effects on haematological and clinical-chemistry parameters were observed in both sexes at the middle and high concentrations. Histologically, effects on the respiratory tract, spleen, thymus, testes and endometrium were found down to the lowest concentration. A no effect level has not been reported.

In rabbits, the chemical has a corrosive effect on the skin and a slight irritant effect on the eye, and in guinea-pigs it induces skin sensitization.

Repeated oral administration of p-chlorobenzotrichloride to rats for 90 days causes inhibition of body weight gain and reductions in the leukocyte and lymphocyte counts in the blood at doses of 12.5 and 25 mg/kg body weight. Histologically, there are effects on the testes and liver. The no effect level is given as 1.25 mg/kg body weight.

In rats and guinea-pigs exposed repeatedly by inhalation for 4 months, weight loss and functional disturbances of the nervous system, liver, kidneys, lungs and blood occur. After exposure to 9.67 mg/m^3 air, morphological changes have been identified in the lungs, in the nervous system and in the liver, sometimes with lethal consequences after 1 month. The medium and low concentrations (1.72 and 0.1 mg/m^3 air, respectively) lead intermittently to functional disturbances (no further details).

The chemical has a mutagenic effect in the Salmonella/microsome test in strain TA 98, and also to a lesser extent in strains TA 100 and TA 1537, independent of whether S9-mix is added or not. A mutagenic effect is also evident in the HPRT test in Chinese hamster V79 cells in vitro.

On repeated oral or dermal exposure in long-term tests in mice, a dose-dependent increase in the tumour incidence is seen in various organs and tissues.

No embryotoxic or teratogenic effects have been found after inhalation exposure of pregnant rats to 4, 10 or 25 mg/m^3 from day 6 to day 19 of pregnancy.

2.   Name of substance

2.1   Usual name   p-Chlorobenzotrichloride

2.2   IUPAC-name $\alpha,\alpha,\alpha,4$-Tetrachlorotoluene

2.3	CAS-No.	5216–25–1
2.4	EINECS-No.	226–009–1

3.   Synonyms, common and trade names

p-Chlorbenzotrichlorid
4-Chlorobenzotrichloride
p-Chlorophenyltrichloromethane
p-Chloro-α,α,α-trichlorotoluene
1-Chloro-4-(trichloromethyl)-
    benzene
4-Chloro-1-(trichloromethyl)-
    benzene
4-Chloro-1-trichloro-methyl-
    benzene
α,α,α,p-Tetrachlorotoluene
p,α,α,α-Tetrachlorotoluene
4,α,α,α-Tetrachlorotoluene
α,α,α-Trichloro-4-chlorotoluene
p-(Trichloromethyl)-chloro-
    benzene

4.   Structural and molecular formulae

4.1   Structural formula

4.2   Molecular formula $C_7H_4Cl_4$

5.   Physical and chemical properties

5.1   Molecular mass, g/mol   229.9

5.2   Melting point, °C   4–5 (Hoechst, 1991)

5.3	Boiling point, °C	240 (Hoechst, 1991)
5.4	Density, g/cm^3	1.48 (at 20 °C) (Hoechst, 1991)
5.5	Vapour pressure, hPa	0.04 (at 20 °C) (Hoechst, 1991)
5.6	Solubility in water	Dissolves with difficulty; slowly hydrolyses (Hoechst, 1989)
5.7	Solubility in organic solvents	Soluble in alcohol, ether, acetone (CHIP, 1983)
5.8	Solubility in fat	No information available
5.9	pH-value	No information available
5.10	Conversion factor	1 ml/m^3 (ppm) ≙ 9.38 mg/m^3 1 mg/m^3 ≙ 0.11 ml/m^3 (ppm) (at 25 °C and 1013 hPa)

6. Uses

Intermediate in the manufacture of herbicides, pharmaceuticals, dyestuffs and disinfectants (Chalepo et al., 1984; Quistad et al., 1985; Hoechst, 1991).

7. Experimental results

7.1 Toxicokinetics and metabolism

[14]C-Ring-labelled p-chlorobenzotrichloride was administered in rapeseed oil to fasted female Sprague-Dawley rats (weight 166 to 186 g) as a single dose by gavage. The doses used were 1.5 mg/kg body weight (2 animals, administered radioactivity ca. 5 μCi) and 102 mg/kg body weight (1 animal, administered radioactivity ca. 38 μCi). The animals were kept in metabolic cages for 4 to 6 days. The urine, faeces and CO_2 from the exhaled air were collected daily. The animals were killed after 4 or 6 days. After appropriate processing of the urine, faeces and tissues, the metabolites were quantitatively and qualitatively determined. The results are presented in Tables 1, 2 and 3.

Table 1. Balance of radioactivity after administration of a single oral dose of ^{14}C-labelled p-chlorobenzotrichloride to female rats

	% of administered dose	
	1.5 mg/kg (after 6 days)	102 mg/kg (after 4 days)
Urine	87	77
Faeces	9	14
Body	4	4
$^{14}CO_2$	≤0.02	-[a]
Volatile organic matter	≤0.002	-[a]
Total	100	95

[a] not detectable

Table 2. Metabolites detected in the urine of female rats after administration of a single oral dose of ^{14}C-labelled p-chlorobenzotrichloride

	% ^{14}C in the urine				
	1.5 mg/kg				102 mg/kg
	Day 1	Day 2	Day 3	Days 4–6	Day 1
p-Chlorobenzotrichloride	≤0.3	≤0.9	-	-	≤0.4
4-Chlorobenzoic acid	0.7	4	6	9	5
4-Chlorohippuric acid	96	37	27	8	88
Start zone of the thin layer chromatogram	4	59	64	73	3

Table 3. Metabolites detected in the faeces of female rats after administration of a single oral dose of ^{14}C-labelled p-chlorobenzotrichloride

| | % ^{14}C in the faeces | | |
| | 1.5 mg/kg | 102 mg/kg | |
	Day 1[a]	Day 1[b]	Day 2[c]
Acetonitrile extract	28	38	64
p-Chlorobenzotrichloride	≤0.3	1	1
4-Chlorobenzoic acid	0	7	6
4-Chlorohippuric acid	0.8	13	27
$\alpha,\alpha',4,4'$-Tetrachlorostilbene	8	13	13
Remaining solids	72	2	36

[a] 8.3% of the applied dose

[b] 7.8% of the applied dose

[c] 3.9% of the applied dose

It can be concluded that p-chlorobenzotrichloride is mainly eliminated in the urine as 4-chlorohippuric acid. The majority of the radioactivity excreted in the faeces cannot be assigned to any of the metabolites specified (Quistad et al., 1985).

7.2   Acute and subacute toxicity

The acute oral toxicity of p-chlorobenzotrichloride was studied in male Wistar rats (ca. 172 g). The observation period was 14 days. The calculated LD_{50} was 0.46 ml/kg (ca. 680 mg/kg body weight). Signs of toxicity observed included increased diuresis, sedation, weight loss, a deterioration in general condition, prostration and "bloody" eyes. The symptoms occurred 1 hour after administration, were of mild to marked severity and persisted until the last day of the study. Deaths were noted from day 2 to day 7 of the study (Bayer, 1980 a).

Further oral LD_{50} values of 1350 mg/kg body weight for rats and 700 mg/kg body weight for mice have been reported (no further details; Chalepo et al., 1984).

A study of the acute oral toxicity of p-chlorobenzotrichloride (purity at least 97%), which was carried out in accordance with OECD guideline no. 401, gave LD_{50} values of 652 mg/kg body weight for male rats and 581 mg/kg body weight for female rats. An LD_{50} of 614 mg/kg body weight was calculated for both sexes together. Clinical signs of toxicity that occurred on the day of administration included abnormally wide-open eyes, exophthalmus, mydriasis, impaired posture and locomotion and spasmodic breathing. The impairments of breathing, posture and movement were still present 1 to 2 days after administration. In addition, miosis and, at the higher dose levels, weakened reflexes and lacrimation were observed. In the animals that died during the study, macroscopic examination revealed patchy lightening of the liver, bulging fullness of the stomach, no digestive contents in the colon and paleness of the spleen at the high dose levels. There were no macroscopically visible changes in the animals examined at the end of the observation period (Hoechst, 1984).

The acute dermal toxicity of p-chlorobenzotrichloride (97.7% pure) was studied in male and female Wistar rats in accordance with OECD guideline no. 402. An LD_{50} value of 1900 mg/kg body weight was calculated. Deaths occurred for up to 4 days after application. Observed signs of toxicity included impaired breathing, coordination and reflexes. In addition the rats showed splayed legs, partly closed or abnormally wide-open eyes, lacrimation, encrustation of blood on the muzzle and at the edges of the eyelids, tremors and increases in spontaneous activity. Effects seen at the site of application from day 2 after application until the end of the study (14-day observation period) included temporary severe reddening, incrustation, open wounds, scab formation and sloughing of skin flakes. Post-mortems on the rats that died revealed blood in the small intestine, and spleens that were light in colour. The rats killed at the end of the study were free from macroscopically visible effects (Hoechst, 1988 a).

The LC_{50} after inhalation exposure (duration not specified) was 125 mg/m^3 air in rats and mice (no further details; Chalepo et al., 1984).

In an inhalation hazard test, 200 l air/hour was passed through ca. 50 g p-chlorobenzotrichloride. The vapour-enriched air produced in this way was inhaled by 5 male and 5 female rats for 7 hours (whole body exposure). The observation period was 14 days. The rats tolerated the 7-hour exposure without showing signs of toxicity. No deaths occurred and no effects on the organs were detectable macroscopically (Bayer, 1980 b).

In a range-finding study, groups of 6 male and 6 female Sprague-Dawley rats (males 165 to 205 g, females 120 to 146 g) were treated with p-chlorobenzotrichloride in corn oil by gavage on 14 consecutive days at doses of 0 (control), 25, 75, 150 and 300 mg/kg body weight/day (experiment 1). Because the compound was toxic at all doses tested in experiment 1, two lower doses (1.25 and 12.5 mg/kg body weight/day) were used in a supplementary study (experiment 2). No clinical signs of toxicity were found on administration of 1.25 mg/kg, while at 12.5 mg/kg the fur was occasionally smeared with faeces and ruffled. At doses of 25 to 300 mg/kg, reduced body weight gain and lower feed intake, body weight loss, gastrointestinal disturbances, a general deterioration in condition, dehydration, impaired breathing, ataxia and spasms occurred dose-dependently. At the highest dose level (300 mg/kg) all of the rats died within 3 to 8 days, while at the next highest level (150 mg/kg), 2 of 12 died. The clinical signs of toxicity were less marked during the second week of the study. Macroscopic effects in the rats that died during the study included red foci in the stomach and pale coloured small intestines, with 2 female rats having adrenal glands with a soft consistency and 1 male having dark-red coloured lungs. At 150 mg/kg an increase in the absolute and relative weights of the adrenal glands was found in both sexes, and the fur was urine-stained. 3 males had small testes and 1 male had chromodacryorrhoea. 1 female showed emaciation and a red-stained muzzle. No treatment-related macroscopic effects were evident in the males that received 75 or 25 mg/kg, while one female in the 75 mg/kg group had a dry red-coloured nose and muzzle and one female in the 25 mg/kg group had a brown-coloured liver. Rats that were given 12.5 or 1.25 mg/kg showed no treatment-related macroscopic effects. Based on these results, doses of 25, 12.5 and 1.25 mg/kg body weight/day were chosen for a 90-day study. The no effect level under the study conditions specified was 1.25 mg/kg body weight/day (see section 7.5; Springborn Laboratories, 1989 a).

Groups of 10 male and 10 female albino rats (Crl:(WJ)BR; weight at the beginning of the study 90 to 115 g) inhaled p-chlorobenzotrichloride in vapour form (whole body exposure) for 6 hours/day, 5 days/week, for 30 days at concentrations of 0 (controls), 4, 20 and 100 mg/m^3. During the study, 3 rats died at the highest concentration (2 males and 1 female). Clinical signs of toxicity observed during exposure at the highest concentration included closed eyes, abnormal posture, sneezing, increased cleaning behaviour, irregular and laboured breathing and wheezing. At the remaining observation times, red coloration around the muzzle and eyes, hair loss around the muzzle, and gasping were evident. At the highest concentration the rats lost weight and feed and water intake were significantly reduced. In the males at the intermediate and high concentrations the haemoglobin level, number of erythrocytes and the average haemoglobin concentration of the individual erythrocytes were increased at the end of the study, while the haematocrit value was also increased at the highest concentration. The average individual erythrocyte volume was significantly decreased in male rats at 100 mg/m^3 as was the number of erythrocytes in females at 100 and 20 mg/m^3. The total number of white blood corpuscles was significantly decreased at the highest concentration due to a significantly reduced lymphocyte count. Clinical chemistry studies revealed reduced albumin, creatinine, transaminase and calcium levels in the males at the highest concentration and reduced transaminase and albumin levels in those at the intermediate concentration. In females at the highest concentration, the phosphorus level was significantly increased and the cholesterol level significantly reduced. The volume of urine was lower in the rats in the 100 mg/m^3 group than in the controls (significantly in the males), an effect that was attributable to reduced water consumption. Macroscopically, the thymus and testes were decreased in size at the highest concentration. The skin of the females showed ulceration and scab formation, and that of the males alopecia. The weights of the pituitary gland, heart, spleen, thymus, thyroid gland and testes in the males, and of the brain and uterus in the females were significantly reduced. Bone marrow myelograms showed a significant reduction in the total number of nucleated cells. In the females, the proportion of basophils was significantly higher. Microscopically, there were dose-dependent effects on the airways, such as atrophy of the olfactory epithelium, and ulceration, squamous metaplasia and hyperplasia of the respiratory epithelium. These effects were pronounced in all rats at the highest concentration level, but were also observed in some rats at the intermediate and low concentrations. In addition, a reduced cell count in the red and white pulp of the spleen, thymic involution, testicular atrophy and reduced endometrial thickness were found at the highest concentration. The effects seen in the spleen correlated with the reduction in the number of circulating lymphocytes. A no effect level could not be determined from these results (HRC, 1984 a).

7.3   Skin and mucous membrane effects

In a patch test, 500 µl of the liquid chemical was applied occlusively to the shaved flank skin of 6 male albino rabbits (New Zealand white, weight 2.14 kg) for 24 hours. Within 24 hours the skin showed a white coloration and a superficial, whitish scab had formed over the treated area. After 8 days the skin was dried out and slightly brown. The chemical was thus corrosive to rabbit skin (FhG, 1980 a).

The substance caused severe local skin irritation in the rabbit. The effect was described as dermatitis, which cleared up only slowly (no further details; Chalepo et al., 1984).

To test the irritant effect on the eye, 100 µl of the liquid chemical was instilled into the right conjunctival sac of each of 6 male albino rabbits (New Zealand white, weight 2.1 kg) as a single dose. The untreated left eye served as a control. The eyes were examined 24, 48 and 72 hours and 8 days after the start of the study. Slight cloudiness of the cornea was seen from 24 to 72 hours after instillation. The iris was not affected. The conjunctiva was slightly red and slight chemosis and secretion were observed, which cleared up during the 8-day observation period. The substance was thus slightly irritating to the eye (FhG, 1980 b).

A slight irritant effect was seen on direct application of p-chlorobenzotrichloride to the cornea in rabbits (no further details on the study protocol; Chalepo et al., 1984).

7.4   Sensitization

The chemical was shown to have clear skin sensitizing potential in an epicutaneous test in guinea-pigs. Induction consisted of the application of a 1% solution to the skin 10 times, followed by challenge with a 1% solution, which caused skin-sensitization effects in 60% of the animals (no further details of protocol; Chalepo et al., 1984).

7.5   Subchronic and chronic toxicity

In a subchronic toxicity study, groups of 10 male and 10 female Sprague-Dawley rats (males 154 to 212 g, females 120 to 170 g) were given p-chlorobenzotrichloride in corn oil orally for 90 days at doses of 1.25, 12.5 and 25 mg/kg body weight/day. The controls were treated with the vehicle. An ophthalmic examination was carried out on all rats at the beginning and end of the study. No deaths occurred during the study. Clinical signs of toxicity observed after administration included salivation and urine-stained fur in both sexes at the top dose and in males at the middle dose, as well as ruffled, faeces-smeared fur in males at the top dose. Body weight gain was reduced in both sexes at the intermediate and top doses. The haematological studies showed a significant reduction in the leukocyte and lymphocyte counts in both sexes at the top dose and in the males at the intermediate dose. The erythrocyte count and haematocrit value in males at the top dose were slightly but significantly lower than those in the control group. No differences from the controls were found with respect to the clinical chemistry studies. Macroscopically, the testes were reduced in size and softer in the high and intermediate dose groups. The absolute and relative testes weights were dose-dependently reduced. Effects correlating with these findings included atrophy of the seminiferous tubules and aspermatogenesis. Aspermia of the epididymides occurred as a secondary effect of aspermatogenesis. Histopathological effects in the females at the top dose were characterized by effects on the liver such as enlarged hepatocytes and vacuolization. The remaining organs showed no pathological effects. Based on these results, a no effect level of 1.25 mg/kg body weight/day was reported (Springborn Laboratories, 1989 b).

In a further, inadequately reported, subchronic inhalation study that was initially planned to last 4 months, rats and guinea-pigs were exposed to analytical concentrations of 9.67, 1.72 and 0.1 mg p-chlorobenzotrichloride/m^3 air (daily duration of exposure not specified). Repeated exposure to 9.67 mg/m^3 air caused pronounced emaciation and marked functional disturbances of the nervous system, liver, kidneys and blood in the rats after 1 month (no further details). Morphological changes were seen in the lungs (purulent destructive bronchitis, focal pneumonia and alveolar bleeding), in the nervous system (pyknosis of the large neurons in the cerebral cortex, isolated neurons with central chromatolysis and vacuolization of the cytoplasm) and in the liver (anisocaryocytosis of the hepatocytes and slight, focal leukocyte infiltration). Because some of the animals died, the study was discontinued (time point not specified). The concentration of 1.72 mg/m^3 air caused temporary body weight reduction and affected the nervous system (phasic effects on the functional condition of the central and peripheral systems). Morphological changes to the cerebral cortex and the brain stem were moderate and transitory (not further described). Effects on the respiratory system included a decrease in breathing frequency and a reduction in the minute volume. There were slight morphological changes in the lungs (moderate hyperplasia of the lymphatic system in the lumen of the bronchi and the vessels, indications of hypertrophy of the bronchial epithelia, an increase in the diameter of the bronchioli and focal emphysema). During a recovery period (duration not specified) the adverse effects proved to be reversible. Repeated exposure to 0.1 mg/m^3 air led to temporary functional disturbances in the rats (not further described), but no morphological effects. Based on the comparison of a limited number of parameters (body weight, blood, kidney function), guinea-pigs were less sensitive. At 1.72 mg/m^3, p-chlorobenzotrichloride caused a reduction in body weight, an increase in the protein content of the urine, and a temporary fall in haemoglobin content and leukocyte count. No conspicuous adverse effects were observed at 0.1 mg/m^3 (no further details; Chalepo et al., 1984). Based on the findings in the lungs, chronic respiratory disease can be assumed to have been present in the rats. The value of this study should therefore be regarded as limited.

7.6   Genotoxicity

7.6.1 In vitro

p-Chlorobenzotrichloride (>97% pure) was tested for mutagenic effects in the *Salmonella typhimurium* strains TA 98, TA 100, TA 1535, TA 1537 and TA 1538 and in *Escherichia coli* WP2uvrA, with and without the addition of a metabolizing system (liver homogenate from Aroclor 1254-treated rats). Based on preliminary bacteriotoxicity studies, 7 different concentrations ranging from 0.16 to 500 µg/plate were tested. A mutagenic effect was shown in strain TA 98 both with and without metabolic activation. A slight increase in the revertant count was also found in strains TA 100 (1.6- to 2.1-fold) and TA 1537 (2.0- to 2.5-fold) both with and without metabolic activation. Thus p-chlorobenzotrichloride induced mutations in this test system (Hoechst, 1986).

p-Chlorobenzotrichloride (minimum 97% pure) was also found to have a mutagenic effect in an HPRT test in Chinese hamster V79 cells in vitro, with and without metabolic activation (S9-mix from

Aroclor 1254-induced rat liver). The concentrations used were 100, 200, 300, and 500 µg/ml without metabolic activation, and 100, 300, 500, 750 and 850 µg/ml with metabolic activation. In the presence of a metabolic activation system, dose-dependent significant increases in the number of mutated colonies and the mutation frequency were seen at 500 µg/ml and above. No induction of mutagenic effects was evident without metabolic activation (Hoechst, 1988 b).

7.6.2 In vivo

No information available.

7.7   Carcinogenicity

Groups of 30 female mice (8 weeks old, ICR strain) were given p-chlorobenzotrichloride by stomach tube twice weekly for 17.5 weeks. The following doses were administered in 0.1 ml sesame oil: 0.05, 0.13, 0.32, 0.8 and 2 ml/mouse. The entire experiment lasted 18 months. The number and distribution of observed tumours is shown in Table 4:

Table 4. Tumour spectrum in female mice following oral administration of p-chlorobenzotrichloride for 17.5 weeks

Group Dose (µl/animal)	A 2	B 0.8	C 0.32	D 0.13	E 0.05	Control
Number of animals/group	29	29	22	28	22	26
Average survival time*	6.2	14.8	16.9	16.9	17.9	17.5
Tumours						
Forestomach						
Squamous epithelial carcinomas	7	6	–	–	–	–
Carcinomas in situ	3	4	1	–	–	–
Multiple papillomas	1	2	5	4	2	–
Glandular stomach carcinomas	–	–	–	–	1	–
Lungs						
Adenocarcinomas	2	15	10	7	3	–
Multiple adenomas	17	10	6	1	2	1
Malignant lymphomas	5	–	–	1	–	1
Thymomas	8	4	–	–	–	–
Skin tumours	6[1]	2[2,3]	1[2]	–	–	–
Other tumours	1[4]	3[5]	1[6]	1[7]	–	–
Number of animals with tumours						
Malignant	16	20	10	8	4	1
Benign	9	7	7	2	2	1
Total	25/29	27/29	17/22	10/28	6/22	2/26

[1] Squamous epithelial carcinoma, [2] Spindle cell sarcoma, [3] Sweat gland carcinoma, [4] Mammary adenocarcinoma, [5] Squamous epithelial carcinoma in the auditory canal, 2 salivary gland adenocarcinomas, [6] Granulosa cell tumour of the ovary, [7] Benign stomach cell carcinoma in situ, * Average survival time is expressed in months after the end of administration

The survival time in the highest dose group was substantially short-ened (6.2 months compared with 17.5 months in the controls). The first tumours observed were thymus tumours and malignant lym-phoadenomas (from month 4). From about month 5, squamous epi-thelial carcinomas developed in the forestomach. Lung tumours occurred most frequently. In addition, neoplastic changes to the skin, breast and salivary glands were observed. Thus, according to the interpretation of the authors, p-chlorobenzotrichloride shows carcinogenic activity after oral administration (Fukuda et al., 1979).

p-Chlorobenzotrichloride (5 µl/mouse) in benzene was applied to the shaved dorsal skin (shoulder blade) of 22 female mice (7 weeks old, ICR-SLC strain) twice weekly for 30 weeks. The mice were observed for up to 9 months after the beginning of application. A control group of 20 female mice received weekly applications of 25 µl benzene/mouse in the same manner as the treated animals. The results are shown in Table 5:

Table 5. Tumour spectrum following application of p-chlorobenzo-trichloride (A) and benzene (control) (B) to the skin for 30 weeks

Substance	A	B
Dose/mouse	5 µl	25 µl
Number of applications/week	2	1
Application period (weeks)	30	30
Number of animals	22	20
Study duration (days)	278	551
Tumours		
Skin		
Squamous epithelial carcinoma	12	0
Sarcoma	2	0
Papilloma	2	0
Lung carcinoma	1	0
Oesophageal carcinoma	5	0
Forestomach carcinoma	2	0
Leukaemia	1	0
Thymus tumours	1	0
Number of animals with tumours		
Malignant	16	0
Benign	2	0
Total	18/22	0/20

The authors considered that p-chlorobenzotrichloride had a tumor-igenic effect following dermal application (Matsushita et al., 1977; Fukuda et al., 1979). However, only one sex, one dose group and 22 animals were tested in this study. No details of non-neoplastic effects are given and there is no exact description of the methods.

7.8   Reproductive toxicity

Groups of 25 pregnant albino rats (Crl:COBS CD(SD)BR strain) inhaled p-chlorobenzotrichloride vapour at concentrations of 0, 4, 10 and 25 mg/m^3 for 6 hours daily from day 6 to day 19 of pregnancy. The rats were killed on day 20, the litter size determined and the foetuses examined for visceral and skeletal anomalies. Feed consumption and body weight gain were reduced in the dams at the highest concentration. No deaths or clinical signs of toxicity were observed. Litter size and pre- and post-implantation losses were not affected in any of the treated groups compared with the control, although the average foetal weight was reduced at the highest concentration. An increase in the incidence of foetuses with cervical ribs (rudimentary unilateral or bilateral ribs on the 7th cervical vertebra) and with non-ossified sternebrae was seen in the foetuses at the highest concentration, effects which were however still within the range of historical control values. The increased number of foetuses with non-ossified sternebrae was related to the low foetal weight at the maternally toxic dose. No teratogenic effects were seen (HRC, 1984 b).

7.9   Effects on the immune system

No information available.

7.10 Neurotoxicity

No information available.

7.11 Other effects

In the Lim-test (short-term test for cumulative effects) a factor of 1.6 was reported following repeated administration by stomach tube.

The authors interpreted this as indicating that the chemical had a marked cumulative effect (no further details; Chalepo et al., 1984).

The same group of authors reported that the chemical was clearly absorbed through the skin. The time span until the onset of death (TL_{50}), which was measured in the muscles, was 2 hours (no further details; Chalepo et al., 1984).

It is not possible to evaluate these studies, due to lack of data.

8.   Experience in humans

No information available.

9.   Threshold limit values

A workplace tolerance value of 0.01 mg/m^3 air has been suggested for the former USSR, with the addendum "penetrates the skin" (Chalepo et al., 1984).

References

Bayer AG, Institut für Toxikologie
p-Chlorbenzotrichlorid 99 destl. - Untersuchung zur akuten oralen Toxizität an männlichen Wistar-Ratten
Unpublished report (1980 a)

Bayer AG, Institut für Toxikologie
p-Chlorbenzotrichlorid - Gewerbetoxikologische Untersuchungen
Unpublished report no. 9063 (1980 b)

Chalepo, A.I., Veselovskaja, K.A., Lapina, L.M., Popova, S.M., Voroncov, R.S.
Materialien zur Begründung der maximal zulässigen Konzentration von α,α,α-Trichlor-4-chlortoluol in der Luft am Arbeitsplatz (German translation from the Russian)
Gig. Tr. Prof. Zabol., 6, 41–43 (1984)

CHIP (Chemical Hazard Information Profile)
p-Chlorobenzotrichloride, draft report (1983)
Environmental Protection Agency, Office of Toxic Substances

FhG (Fraunhofer-Institut für Toxikologie und Aerosolforschung)
Bericht über die Prüfung von p-Chlorbenzotrichlorid 99 auf primäre
Hautreizwirkung
Unpublished report (1980 a)
On behalf of Bayer AG

FhG (Fraunhofer-Institut für Toxikologie und Aerosolforschung)
Bericht über die Prüfung von p-Chlorbenzotrichlorid 99 auf
Schleimhautreizwirkung
Unpublished report (1980 b)
On behalf of Bayer AG

Fukuda, K., Matsushita, H., Takemoto, K.
Carcinogenicity of p-chlorobenzotrichloride
52nd Ann. Meeting, Japan Assoc. Ind. Health, 330 ff. (1979)
and English translation of the original report, communication from
Hoechst Japan Ltd. (1988)

Hoechst AG, Pharma Forschung Toxikologie
p-Chlorbenzotrichlorid - Prüfung der akuten oralen Toxizität an der
männlichen und weiblichen Wistar-Ratte
Unpublished report no. 84.0041 (1984)

Hoechst AG, Pharma Research Toxicology
p-Chlorbenzotrichlorid - Study of the mutagenic potential in strains
of *Salmonella typhimurium* (Ames Test) and *Escherichia coli*
Unpublished report no. 86.1462 (1986)

Hoechst AG, Pharma Forschung Toxikologie und Pathologie
p-Chlorbenzotrichlorid TTR - Prüfung der akuten dermalen
Toxizität an der Wistar-Ratte
Unpublished report no. 88.0121 (1988 a)

Hoechst AG, Pharma Research Toxicology and Pathology
p-Chlorbenzotrichlorid TTR - Detection of gene mutations in somat-

ic cells in culture, HGPRT-test with V79 cells
Unpublished report no. 88.0932 (1988 b)

Hoechst AG
Data sheet „Altstoffe" p-Chlorbenzotrichlorid (1989)

Hoechst AG
AIDA-Grunddatensatz: Benzene, 1-chloro-4-(trichloromethyl)-(9CI)
(1991)

HRC (Huntingdon Research Centre, Ltd.)
Para-chlorobenzotrichloride vapour thirty day inhalation toxicity
study in rats
Report no. HKR 10/84109 (1984 a)
On behalf of the Occidental Chemical Corporation
NTIS/OTS 0537003

HRC (Huntingdon Research Centre, Ltd.)
Effect of para-chlorobenzotrichloride p-CBTC on pregnancy of the rat
Report no. HKR 13/841215 (1984 b)
On behalf of the Occidental Chemical Corporation
NTIS/OTS 0535896

Matsushita, H., Fukuda, K., Takemoto, K.
Carcinogenicity of benzotrichloride and its related compounds
Unpublished report, National Institute of Industrial Health, Ministry
of Labour, Japan (1977)

Quistad, G.B., Mulholland, K.M., Skiles, G., Jamieson, G.C.
Metabolism of 4-chlorobenzotrichloride in rats
J. Agric. Food Chem., 33 (1), 95–98 (1985)

Springborn Laboratories, Inc., Mammalian Toxicology Division,
Spencerville, Ohio
Two-week range-finding study of parachlorobenzotrichloride in rats
Report of SLS study no. 3073.35 (1989 a)
On behalf of the Occidental Chemical Corporation
NTIS/OTS 0525376 and 0535898

Springborn Laboratories, Inc., Mammalian Toxicology Division, Spencerville, Ohio
90-day oral toxicity study of parachlorobenzotrichloride in rats
Report of SLS study no. 3073.36 (1989 b)
On behalf of the Occidental Chemical Corporation
NTIS/OTS 0526376 and 0535897

Compound chemical index vol. 1-9

P-Chlorophenyltrichloro-
methane *9*, 139
C.I. 37030 *2*, 88
Clorox *4*, 259
CNA *1*, 75
m-Cresidine *1*, 173
Cresol diphenyl phosphate
5, 160
Cresyl diphenyl phosphate
5, 160
Cryolite *5*, 144
Cryolith *5*, 144
p-Cumenyl isocyanate *7*, 193
psi-Cumohydroquinone
5, 126
Cyanuric acid (Monosodium
cyanurate) *7*, 48
Cyanursäure *7*, 49

Daito Red Base RL *2*, 101
Dakins solution *4*, 259
Daltogen *4*, 127
Dan Klorix *4*, 259
Dazzle *4*, 259
DEA *2*, 136
Deactivator E *1*, 219
Deactivator H *1*, 219
DEC *7*, 36
DEG *1*, 219
Deosan *4*, 259
DETA *7*, 116
Devol Red RL *2*, 101
DGE *3*, 156
Diabase Red RL *2*, 101
Diacetic ester *7*, 163
Diafen FF *7*, 2
1,4-Diamino-9,10-anthracene-

dione *4*, 223
1,4-Diaminoanthrachinon
4, 223
1,4-Diaminoanthraquinone
4, 223
1,5-Diaminonaphthalene
2, 192, 193
β,β-Diaminodiethylamine
7, 116
2,2'-Diaminodiethylamine
7, 116
1,4-Dianilinobenzene *7*, 2
Diatol *7*, 36
3,6-Diazaoctane-1,8-diamine
4, 183
Diazenedicarboxamide
6, 134, *8*, 112
Diazenedicarbonic acid amide
6, 134, *8*, 113
Diazendicarbonsäureamid
6, 134, *8*, 113
2',4'-Dichloro-4-aminobiphen-
yl ether *8*, 226
2',4'-Dichloro-4-aminodiphe-
nyl ether *8*, 226
Dichlorodimethylsilane *9*, 2
Dichlorodimethylsilicon *9*, 2
Dichloromethyl benzene
4, 58
4-(2,4-Dichlorophenoxy)ani-
line *8*, 226
p-(2,4-Dichlorophenoxy)ani-
line *8*, 226
4-(2,4-Dichlorophenoxy)-ben-
zeneamine *8*, 225
Dichlorophenyl methane
4, 58
α,α-Dichlorotoluene *4*, 58
α,α-Dichlorotoluol *4*, 58

Phenoxypropene oxide 3, 74
p-Phenylaminodiphenylamine
  7, 2
Phenyl carbonimide 4, 152
Phenyl isocyanate 4, 152
Phenylcarbimide 4, 152
Phenylchloroform 4, 39
Phenylfluorid 8, 192
Phenyl fluoride 8, 192
Phenylfluoroform 7, 27
Phenylglycidyl ether 3, 74
Phenylisocyanat 4, 152
Phenylsilicon trichloride 9, 70
Phenylsilicontrichloride 9, 70
Phenyltrichloromethane 4, 39
Phenyltrichlorosilane 9, 69
Phenyltrimethyl-ammonium
  chloride 1, 311
N-Phenyl-1,4-benzenediamine
  5, 46
N-Phenyl-1,4-benzoldiamin
  5, 46
N-Phenyl-p-phenylenediamine
  5, 46
Phosflex 4 1, 298, 8, 139
Phosflex 112 5, 161
Phosphoric acid, cresyl
  diphenyl ester 5, 161
Phosphoric acid, diphenyl
  tolyl ester 5, 161
Phosphoric acid, methyl-
  phenyl diphenyl ester 5, 161
Phosphoric acid tri-n-butyl-
  ester 1, 298, 8, 138
Phosphorsäurediphenyltolyl-
  ester 5, 161
Phthalic acid dinitrile 2, 76
o-Phthalodinitrile (o-PDN)
  2, 76

Phthalodinitrile 2, 76
Pichtosin 9, 92
PNP 6, 54
Poly-Solv DE 5, 92
Poly-Solv DM 8, 85
Pomarsol 3, 93
Pomasol 3, 93
Porofor ADC 6, 134, 8, 113
Porofor BSH 6, 34
Potassium tetrafluoroborate
  7, 183
Bis(2-propanol)amine 4, 173
Propaldehyde 6, 114
Propanal 6, 113
Propanaldehyde 6, 114
Propargyl alcohol 2, 121
Propargyl chloride 9, 103
Propargylchlorid 9, 103
2-Propenoic acid 2, 42
2-Propensäureamid,
  N-(Hydroxymethyl)-2-methyl-
  7, 176
Propional 6, 114
Propionaldehyd 6, 114
Propionaldehyde 6, 113, 114
Propionic aldehyde 6, 114
Propionaldehyd 6, 114
Propylaldehyde 6, 114
Propylic aldehyde 6, 114
Propyne-(1)-ol-(3) 2, 121
2-Propyne-1-ol 2, 121
Propyne-(2)-ol 2, 121
1-Propyne-3-ol 2, 121
2-Propynol-1 2, 121
Pseudocumohydroquinone
  5, 126
p-Pseudocumoquinone
  2, 178

Tulabase Fast Red RL   *2*, 102

UN 1181   *5*, 71
Urea, 1,1'-isobutylidenedi-
   6, 158, *8*, 201
Urea, N,N"-(2-methylpropyli-
   dene)bis-   *6*, 158, *8*, 201

Vanlube 81   *5*, 65
VF   *2*, 1
Villaumite   *5*, 144
Vinyl carboxylic acid   *2*, 42
Vinyl fluoride   *2*, *1*, 2
Vinyfor AC   *6*, 134, *8*, 113
Vinyl methyl ether   *5*, 83
Vulcacit NPV/C   *1*, 83
Vulcafor DOTG   *6*, 42
Vulkacit DOTG   *6*, 42
Vulkacit P extra N   *6*, 214
Vulkacit Thiuram   *3*, 93

Wacker Silan P   *9*, 70
Warecure C   *1*, 83

asym. m-Xylidine   *8*, 11
m-Xylidine   *8*, 11
m-4-Xylidine   *8*, 11
meta-Xylidine   *8*, 11
meta-4-Xylidine   *8*, 11
2,4-Xylidin   *8*,11
2,4-Xylidine   *8*, 10
2,4-Xylylamine   *8*,11

Yamada Fast Red RL Base
   2, 102

Zinc ethylphenyl-dithiocarba-
   mate   *6*, 214
Zinkethylphenyldithiocarba-
   mat   *6*, 214

Springer-Verlag
and the Environment

We at Springer-Verlag firmly believe that an international science publisher has a special obligation to the environment, and our corporate policies consistently reflect this conviction.

We also expect our business partners – paper mills, printers, packaging manufacturers, etc. – to commit themselves to using environmentally friendly materials and production processes.

The paper in this book is made from low- or no-chlorine pulp and is acid free, in conformance with international standards for paper permanency.

MIX
Papier aus verantwortungsvollen Quellen
Paper from responsible sources
FSC® C105338

If you have any concerns about our products,
you can contact us on
ProductSafety@springernature.com

In case Publisher is established outside the EU,
the EU authorized representative is:
Springer Nature Customer Service Center GmbH
Europaplatz 3, 69115 Heidelberg, Germany

Printed by Libri Plureos GmbH
in Hamburg, Germany